SCIENCE QUIZ
(National Children's Science Congress)

Nalin Kumar Ramaul*
Pinki Ramaul**

*Dr. Nalin Kumar Ramaul, Government Degree College, Bharali (Anjbhoj), District Sirmaur, Himachal Pradesh, India 173025
**Pinki Ramaul, Secretary, Stellar Society (Registered NGO), Paonta Sahib, District Sirmaur, Himachal Pradesh, India

Published by:

NIK NOS, Friends Enclave, Shubh Khera, Paonta Sahib, District Sirmaur, Himachal Pradesh 173025

Email: pinkiramaul@gmail.com

(Pinki Ramaul, Proprietor)

First Published in 2018

Science Quiz

1. What is the most important stimulant in tea leaves?
   + Caffeine
2. What is the source of turpentine?
   + Pine
3. Which plant hormone controls fruit ripening?
   + Ethylene
4. Which is the largest living bird?
   + Ostrich
5. Body cells infected with virus produce a protein called?
   + Interferon
6. What is the naturally occurring anti-coagulant in the human blood?
   + Heparin
7. Common salt is obtained from sea water by the process of:
   + Evaporation
8. Washing soda is:
   + Hydrated sodium carbonate
9. Chemically baking soda is:
   + Sodium Bicarbonate
10. Lime water contains:
    + Calcium hydroxide
11. Cooking oil can be converted into vegetable ghee by the process of
    + Hydrogenation
12. Diamond is chemically:
    + Pure Carbon
13. Commercial cork is obtained from the bark of:
    + Oak
14. A plant bearing both male and female flowers is called:
    + Monoecious
15. For how much time can bacteria remain alive at sub-zero temperatures?
    + More than 1000 years
16. The term 'gene' for the factor controlling heredity was coined by:

Science Quiz

    ✦ W. Johannsen
17. How many pairs of chromosomes are there in humans?
    ✦ 23 pairs
18. The 'one gene one enzyme' hypothesis suggesting that one gene controls the synthesis of one enzyme was proposed by:
    ✦ Beadle and Tatum
19. Body cells infected with virus produce a protein called:
    ✦ Interferon
20. Chocolate can be bad for health because of a high content of:
    ✦ Nickle
21. Montreal Protocol is related to:
    ✦ Ozone depletion
22. The nearest planet to the sun is:
    ✦ Mercury
23. The biggest planet of the solar system is:
    ✦ Jupiter
24. The brightest planet is:
    ✦ Venus
25. What is the smallest continent of earth?
    ✦ Australia
26. National Animal:
    ✦ The Tiger (*Panthera tigris*) is the national animal of India.
    ✦ Lion was the national animal of India till 1972. Later on, it was replaced by Tiger.
27. National Bird:
    ✦ The Indian Peacock (*Pavo cristatus*) is the National bird of India.
28. National Flower:
    ✦ Lotus is the national flower of India.
29. National Fruit:
    ✦ Mango is the national fruit of India.
30. National Tree:

Science Quiz

- ✦ Banyan (*Ficus Bengalensis*)
31. National River:
    - ✦ The Ganga
    - ✦ It is the longest river of India.
    - ✦ It originates in the snow fields of the Gangotri glacier in the Himalayas as the Bhagirathi river.
32. National Aquatic Animal:
    - ✦ The Ganges River Dolphin was declared as National Aquatic Animal in October 2009.
33. National Heritage Animal:
    - ✦ In October 2010, the elephant was declared the national heritage animal of India.
34. Scalar Quantity:
    - ✦ Which has only magnitude
35. Vector Quantity:
    - ✦ Which has both magnitude and direction
36. Who initially gave the idea of laser?
    - ✦ Albert Einstein
37. Who is honoured as Father of Modern Chemistry?
    - ✦ Antoine Lavoisier
38. Who invented periodic table?
    - ✦ Dmitri Mendeleev
39. Which is the most abundant gas in the earth's atmosphere?
    - ✦ Nitrogen (78.08%)
40. Which gas evolved from paddy fields and marshes?
    - ✦ Methane
41. Which toxic element present in automobile exhausts?
    - ✦ Lead
42. Which gases cause acid rain?
    - ✦ Sulphur dioxide, Nitrogen oxides
43. Which metal is used in the making of microchips?
    - ✦ Silicon
44. What is the toxicity caused by silicon called?
    - ✦ Silicosis
45. Which polymer is used to manufacture electric switches,

computer disks etc.?
- ✦ Bakelite
46. With which polymer, the cabinets of radio and TV made of?
    - ✦ Polystyrene
47. Which chemical is used to make rain coats?
    - ✦ Poly Vinyl Chloride (PVC)
48. Which type of plastics can be recycled?
    - ✦ Thermoplastics
49. What is the name given to plastics that cannot be recycled?
    - ✦ Thermosetting plastics
50. Which element is a liquid at ordinary temperature?
    - ✦ Mercury
51. What is the chemical name of baking soda?
    - ✦ Sodium bicarbonate
52. What is sodium carbonate commonly known as?
    - ✦ Washing soda or soda ash
53. What is sodium chloride NaCl?
    - ✦ Table salt
54. What is table sugar?
    - ✦ Cane sugar (Sucrose)
55. What is the element present on lead pencils?
    - ✦ Carbon (graphite)
56. Which gas is found in soda water?
    - ✦ Carbon dioxide?
57. What is the solvent of gold?
    - ✦ Aqua Regia
58. Name the chemical assed to make tooth pastes white?
    - ✦ Titanium dioxide
59. Which chemical is the reason behind the brown colour of human faces?
    - ✦ Bilirubin
60. Which chemical is called Chile Salt peter?
    - ✦ Sodium Nitrate
61. What is marsh gas?
    - ✦ Methane

Science Quiz

62. What is the chemical name of Chinese salt?
    - Monosodium glutamate
63. Which is the purest form of carbon?
    - Graphite
64. Which is called white poison?
    - Sugar
65. Which is the simplest sugar?
    - Glucose
66. What is the biological polymer in paper?
    - Cellulose
67. What is the chemical name of Vitamin E?
    - Alpha tocopherol
68. Which Vitamin is also known as retinol?
    - Vitamin A
69. What is Calciferol?
    - Vitamin D
70. What is the other name of thiamine?
    - Vitamin B1
71. Which Vitamin is called niacin?
    - Vitamin B3
72. Which element is excreted through human sweat?
    - Sulphur
73. Pollution of water bodies by which element causes Minamata disease?
    - Mercury
74. Accumulation of which chemical leads to kidney stones?
    - Calcium Oxalate
75. Which heavy metal causes Itai Itai disease?
    - Cadmium
76. Which chemical was used earlier for giving anaesthesia to patients?
    - Diethyl ether
77. The Indian Space Research Organization (ISRO) Navigation centre was set up at
    - Byalalu, about 40 km from Bangalore

78. The acronym _____ means billion of operations per second
    ✦ GB
79. A technology to provide high speed internet access over telephone wiring:
    ✦ ADSL
80. Which device is used to study the way and object behaves when the air flows over it?
    ✦ Wind Tunnelp
81. The part of a computer that shows visual information on a screen.
    ✦ Visual Display Unit (VDU)
82. What are the machines used to check the progress of unborn babies still in the mother's womb?
    ✦ Ultra Sound Machines
83. A type glass that is highly resistant to heat.
    ✦ Borosilicate Glass
84. The process of killing diseases producing microorganism in food items by heat.
    ✦ Pasteurization
85. Tuberculosis is caused by it.
    ✦ Bacteria
86. Which country in the world has maximum number of robots working?
    ✦ Japan
87. What is the name of the instrument that measures wind speed?
    ✦ Anemometer
88. A device used to measure the strength of magnetic field.
    ✦ Magnetometer
89. A strip of flexible plastic with magnetic coating that is used to record sound and pictures.
    ✦ Magnetic Tape
90. A method of growing plants without soil.
    ✦ Hydroponics

Science Quiz

91. Electricity produced from falling water.
    + Hydroelectricity
92. The fear of being out of mobile phone contact is known as?
    + Nomophobia
93. In which year did IT Act come into force in India?
    + 2000
94. A new technology which provides the ability to create an artificial world and have people interact with it is called?
    + Televirtuality
95. The device used for locating submerged object sunder sea is-
    + Sonar
96. A baby blue whale drinks this many litters of milk per day:
    + 190
97. For what is the Jurassic period named?
    + mountain range in Switzerland
98. Which blood group is called the "universal donor"?
    + O
99. A banana plant fits into which of these groups?
    + Herbs
100. From what plant was paper originally made?
    + Papyrus (The ancient Egyptians made a parchmentlike paper, also called papyrus, from the pith, or inner portion, of the stalk of the papyrus plant.)
101. What is a dehiscent plant?
    + The fruit of a dehiscent plant splits open when it is ripe to disperse seeds. Examples are peas, beans, peanuts, and magnolias. Technically, a banana is an herb.
102. How much blood is in the human body?
    + 5.7 litters (The average human carries about 192.7 ounces (5.7 litters) of blood. Altogether, this quantity of blood circulates through the body twice a minute.)
103. How many facial muscles are involved in speaking?
    + 70 (About 70 muscles are involved in speaking. These are located in several parts of the head, neck, and

104. How many taste buds do young people have
     + 100,000
105. How many basic tastes can humans sense?
     + 5 (Humans can sense five basic tastes: salty, sweet, sour, bitter, and umami (representing chemicals known as amino acids).
106. What is the word for a person with very pale skin and eyes?
     + Albino
107. What sensory function do the ears provide other than hearing?
     + balance
108. Which of these joints is not found in the human body?
     + mortise and tenon
109. How many sweat glands does the average person have?
     + 2.6 million
110. What scientist has popularized the concept of biodiversity?
     + E.O. Wilson
111. The author of On the Origin of Species is:
     + Charles Darwin
112. Which group of animals is Charles Darwin best known for studying?
     + Birds
113. Which of the following is a poisonous fish?
     + Lion fish (Lion fish are noted for their venomous fin spines, which are capable of producing painful, though rarely fatal, puncture wounds.)
114. What is the study of animal life called?
     + Zoology
115. In science, what is the name for the classification of plants and animals?
     + Taxonomy
116. How often, approximately, is Halley's Comet visible from Earth (in years)?

- ✦ 75 (Halley's comet appears in the night sky approximately every 75 years. Its next appearance will be in 2061.)
117. What astronomer placed Earth at the centre of the universe?
    - ✦ Ptolemy
118. Which astronomer wrote the Aryabhatiya?
    - ✦ Aryabhata I
119. What astronomer suggested that the Sun was at the centre of the solar system?
    - ✦ Copernicus (In 'On the Revolutions of the Celestial Orb', Nicholas Copernicus (1473–1543) suggested that Earth revolved around the Sun.)
120. Who was the first astronomer to use a telescope?
    - ✦ Galileo
121. What do most asteroids orbit?
    - ✦ Sun
122. Which of these planets has an atmosphere made up of hydrogen, helium, and methane?
    - ✦ Uranus
123. Which astronomers discovered Pluto in 1930?
    - ✦ Clyde Tombaugh
124. About how many galaxies are there in the universe?
    - ✦ 25 billion
125. How many light-years across is the Milky Way?
    - ✦ 100000
126. Which is the closest galaxy to Earth outside the Milky Way?
    - ✦ Andromeda
127. The first Indian Remote Sensing Satellite was launched into space on
    - ✦ 1990
128. The code name of the first vehicle driven by man on the moon's surface was
    - ✦ Rover
129. India's satellite launch-pad is located at

- Shri Hari Kota
130. The first country to send man to the moon
- U.S.A.
131. The first person in the world to land on the moon
- Neil A. Armstrong and Edwin E. Aldrin Jr. of U.S.A.
- Armstrong was the first to set foot on the moon followed by Aldrin. July 21, 1969.
132. The first spaceship which carried three American astronauts to land two of them on the moon
- Apollo-11
133. The first man to enter space
- Major Yuri Gagarin (Russian)
134. The first woman cosmonaut of the world
- Velentina Tereshkova
135. The first American astronaut to float in space
- Edward White
136. First crew transfer between the orbiting spaceships
- Soyuz T-15 with Mir Space Station
137. India's first scientific satellite
- Aryabhatta
138. The first Indian to go into space
- Rakesh Sharma
139. The first woman of Indian origin in space
- Kalpana Chawla
140. The first space tourist in the world
- Dennis Tito (U.S.A.)
141. The first US space shuttle to explode while returning home killing all the astronauts
- Columbia (February 1, 2003)
142. India's first Mapping Satellite
- CARTOSAT-I (Launched on May 5, 2005)
143. The first space woman to stay for the longest ever Perivale of time in space
- Sunita Williams
144. The first successful moon mission of India

Science Quiz

- ✦ Chandrayan-I (October 22, 2008)
145. What is the biggest planet in our solar system?
     - ✦ Jupiter
146. What is the chemical symbol for the element oxygen?
     - ✦ O
147. What is the 7th element on the periodic table of elements?
     - ✦ Nitrogen
148. Expand DNA
     - ✦ 'Deoxyribonucleic acid'?
149. The highest mountain on earth is?
     - ✦ Mount Everest
150. What is the name of the closest star to the earth?
     - ✦ The Sun
151. What is the name of the element with the chemical symbol 'He'?
     - ✦ Helium
152. Pure water has a pH level of around?
     - ✦ 7
153. The molten rock that comes from a volcano after it has erupted is known as what?
     - ✦ Lava
154. What is the name of the part of the human skeleton which protects our brain?
     - ✦ The skull
155. Is the compound 'HCl' an acid or base?
     - ✦ An acid (hydrochloric acid)
156. How many bones do sharks have in their bodies?
     - ✦ 0
157. What famous scientist was awarded the 1921 Nobel Prize in Physics for his work on theoretical physics?
     - ✦ Albert Einstein
158. What is the fastest land animal in the world?
     - ✦ The cheetah (it can reach speeds of up to 120kph – 75mph).
159. What is the largest land animal in the world?

Science Quiz

- ✦ The elephant - The largest on record weighed around 12,000 kilograms! (26,000 lb).
160. What is the only continent on earth where Giraffes live in the wild?
    - ✦ Africa
161. What type of animal is the largest primate in the world?
    - ✦ The Gorilla
162. What is the tallest animal in the world?
    - ✦ The giraffe - The average height is around 5 metres (16ft) and the tallest on record stood nearly 6 metres (20 ft) tall.
163. True or false? Sound travels faster through water than air?
    - ✦ True
164. Water is made up of what two elements?
    - ✦ Hydrogen and oxygen
165. The biggest ocean on Earth.
    - ✦ Pacific Ocean
166. The solid state of water is known as what?
    - ✦ Ice
167. True or false? Pure water is tasteless.
    - ✦ True
168. Nimbus, cumulus, and stratus are types of what?
    - ✦ Clouds
169. True or false? Water is an example of a chemical element.
    - ✦ False
170. How much of Earth's surface does water cover?
    - ✦ Around 70%
171. At what temperature does water boil?
    - ✦ 100 degrees Celsius (212 degrees Fahrenheit).
172. When water is cooled, does it contract or expand?
    - ✦ Expand
173. Water freezes at what temperature?
    - ✦ 0 °C (32 °F)
174. The deepest point in all of the world's oceans is named what?

- Mariana Trench
175. What is the longest river on Earth?
     - The Nile River
176. True or false? Ice sinks in water.
     - False - It floats
177. A thermometer is a device used to measure what?
     - Temperature
178. What country experiences the most tornadoes?
     - USA
179. True or false? A rainbow is a spectrum of light that appears when the Sun shines onto water droplets in the air.
     - True
180. Earth's recent temperature rises which have been linked to human activity is known as global _____?
     - Warming
181. What is the highest temperature ever recorded on Earth?
     - 57.8 °C (136 °F)
182. Where was the highest recorded temperature on earth of 57.8 °C (136 °F) recorded?
     - Al 'Aziziyah, Libya
183. What are the scientists who study weather called?
     - Meteorologist
184. What is the driest desert on Earth?
     - The Atacama Desert
185. What is the lowest temperature ever recorded on Earth?
     - − 89.2 °C (−128.6 °F)
186. Where was the lowest recorded temperature of − 89.2 °C (−128.6 °F) on earth recorded?
     - Vostok Station, Antarctica
187. Balls or irregular lumps of ice that fall from clouds (often during thunderstorms) are known as what?
     - Hail stones
188. An avalanche features the rapid descent of _____?
     - Snow

Science Quiz

189. What is the most rain fall ever recorded in one year in Cherrapunji, India?
     ✦ 25.4 meters
190. What is the name of a weather instrument used to measure atmospheric pressure?
     ✦ A barometer
191. An anemometer is used to measure what?
     ✦ Wind speed
192. At what temperature is Centigrade equal to Fahrenheit?
     ✦ – 40 degrees
193. Trying to predict the weather is known as weather _____?
     ✦ Forecasting
194. The scientific study of plant life is known as what?
     ✦ Botany
195. The process of plants using energy from sunlight to turn carbon dioxide into food is known as what?
     ✦ Photosynthesis
196. The movement of pollen from the anthers to the stigma of a flower is known as what?
     ✦ Pollination
197. Amber is made from fossilized tree _____?
     ✦ Resin
198. What grain has the highest level of worldwide production?
     ✦ Maize (corn) (Rice is second)
199. A trailing or climbing plant is also known as a _____?
     ✦ Vine
200. What is the name of the world's largest reef system?
     ✦ Great Barrier Reef
201. Do male or female mosquitoes bite people?
     ✦ Female (The female mosquito is the one that bites (males feed on flower nectar). She requires blood to produce eggs. Her mouthparts are constructed so that they pierce the skin, literally sucking the blood out.)
202. When is the Earth Day celebrated?

Science Quiz

✦ April 22
203. What are the 3 R's of recycling?
    ✦ Reduce, reuse, and recycle
204. What famous islands west of Ecuador were extensively studied by Charles Darwin?
    ✦ Galapagos Islands
205. What is the boiling point of water?
    ✦ 100 degrees Celsius (212 degrees Fahrenheit).
206. Heat from the sun gets to the Earth by radiation, conduction, or convention?
    ✦ Radiation
207. What is the freezing temperature of water?
    ✦ 0 degrees Celsius (32 degrees Fahrenheit)
208. Substances that do not conduct heat are known as what?
    ✦ Insulators
209. What is the chemical symbol of gold?
    ✦ Au
210. True or false? Steel is a chemical element.
    ✦ False - Alloy
211. What is the most common metal found on Earth?
    ✦ Iron
212. What three kinds of medals are awarded at the Olympic Games?
    ✦ Gold, silver, and bronze
213. Bronze is made from what two metals?
    ✦ Copper and tin
214. What is the only metal that is liquid at room temperature?
    ✦ Mercury
215. What metal has the chemical symbol Pb?
    ✦ Lead
216. When a solid change to a liquid it is called what?
    ✦ Melting
217. When a gas changes into a liquid it is called what?
    ✦ Condensation
218. What is it called when a solid change directly into a gas?

- ✦ Sublimation
219. What is the charge on Protons?
    - ✦ Positive
220. What is the charge on Electrons?
    - ✦ Negative
221. True or false? Protons and neutrons are part of the nucleus.
    - ✦ True
222. True or false? Nucleon is a collective name for two particles, the neutron and proton.
    - ✦ True
223. Bases change litmus paper to
    - ✦ Blue
224. What is the chemical formula for sulfuric acid:
    - ✦ H_2SO_4
225. Which acid do lemons contain?
    - ✦ Citric acid
226. True or false? An anion is an ion with more electrons than protons, giving it a negative charge.
    - ✦ True
227. True or false? A cation is an ion with more protons than electrons, giving it a positive charge.
    - ✦ True
228. What is the pH of neutral solutions?
    - ✦ 7
229. What is the pH of acid?
    - ✦ Less than 7
230. What is the pH of base?
    - ✦ Above 7
231. Acids change litmus paper to
    - ✦ Red
232. True or false? The word acid comes from the Latin word acidus (meaning sour).
    - ✦ True
233. What is the chemical formula for sodium hydroxide?

✦ NaOH

234. Which acid gives vinegar a sour taste and strong smell?
    ✦ Acetic acid
235. Which acid is found in bee venom?
    ✦ Formic acid
236. Sodium hydroxide is known as
    ✦ A strong base
237. What are the two main metals in the earth's core?
    ✦ Iron and nickel
238. Which is hotter, the centre of the earth or surface of the sun?
    ✦ The centre of the earth
239. What do you call molten rock before it has erupted?
    ✦ Magma
240. The Great Barrier Reef is found off the coast of which country?
    ✦ Australia
241. What do you call a person who studies rocks?
    ✦ A geologist
242. Over a long period of time while under extreme heat and pressure, graphite turns into which precious mineral?
    ✦ Diamond
243. Someone who studies earthquakes is known as a what?
    ✦ Seismologist
244. What is the name of the layer of earth's atmosphere that absorbs the majority of the potentially damaging ultraviolet light from the sun?
    ✦ The ozone layer
245. The mass of the earth is made up mostly of which two elements?
    ✦ Iron (32%) and oxygen (30%)
246. What is the second most common gas found in the air we breathe?
    ✦ Oxygen (21%)
247. In terms of computing, what does CPU stand for?

✦ Central Processing Unit
248. Is the wavelength of infrared light too long or short to be seen by humans?
✦ Long
249. Firefox, Opera, Chrome, Safari and Explorer are types of what?
✦ Web browsers
250. In terms of computing, what does ROM stand for?
✦ Read Only Memory
251. What does the abbreviation WWW stand for?
✦ World Wide Web
252. When light bends as it enters a different medium the process is known as what?
✦ Refraction
253. A magnifying glass is what type of lens?
✦ Convex
254. Electric resistance is typically measured in what units?
✦ Ohms
255. A person who studies physics is known as a?
✦ Physicist
256. Metals expand when heated and do what when cooled?
✦ Contract
257. What is the first name of the famous scientist who gave us Newton's three laws of motion?
✦ Isaac
258. What state of the art computer technology is used to train pilots when wanting to copy the experience of flying an aircraft?
✦ A flight simulator
259. Electric power is typically measured in what units?
✦ Watts
260. The most recognized model of how the universe begun is known as the?
✦ Big bang
261. The wire inside an electric bulb is known as the what?

- Filament
262. What kind of eclipse do we have when the moon is between the sun and the earth?
    - A solar eclipse
263. Conductors have a high or low resistance?
    - Low
264. Electric current is typically measured in what units?
    - Amperes
265. What scientist is well known for his theory of relativity?
    - Albert Einstein
266. Earth is located in which galaxy?
    - The Milky Way galaxy
267. Electric current is measured using what device?
    - Ammeter
268. In terms of electricity, what does DC stand for?
    - Direct current
269. Who introduced the concept of electric fields?
    - Michael Faraday
270. In terms of electricity, what does AC stand for?
    - Alternating current
271. True or false? You can extend battery life by storing batteries at a low temperature.
    - True
272. Which famous scientist introduced the idea of natural selection?
    - Charles Darwin
273. A person who studies biology is known as a?
    - Biologist
274. Botany is the study of?
    - Plants
275. Can frogs live in salt water?
    - No
276. True or false? The common cold is caused by a virus.
    - True

Science Quiz

277. Animals which eat both plants and other animals are known as what?
    - ✦ Omnivores
278. Bacterial infections in humans can be treated with what?
    - ✦ Antibiotics
279. A single piece of coiled DNA is known as a?
    - ✦ Chromosome
280. The area of biology devoted to the study of fungi is known as?
    - ✦ Mycology
281. What is the name of the process used by plants to convert sunlight into food?
    - ✦ Photosynthesis
282. The death of every member of a particular species is known as what?
    - ✦ Extinction
283. The process of pasteurization is named after which famous French microbiologist?
    - ✦ Louis Pasteur
284. A change of the DNA in an organism that results in a new trait is known as a?
    - ✦ Mutation
285. What is the first element on the periodic table?
    - ✦ Hydrogen
286. What is the centre of an atom called?
    - ✦ A nucleus
287. What is the main gas found in the air we breathe?
    - ✦ Nitrogen (around 78%)
288. K is the chemical symbol for which element?
    - ✦ Potassium
289. What orbits the nucleus of an atom?
    - ✦ Electrons
290. True or false? A neutron has no net electric charge.
    - ✦ True

Science Quiz

291. A nuclear reaction where the nucleus of an atom splits into smaller parts is known as nuclear fission or nuclear fusion?
     ✦ Nuclear fission
292. What is H_2O more commonly known as?
     ✦ Water
293. What is the third most common gas found in the air we breathe?
     ✦ Argon (around 1%)
294. What is the name given to substances that are initially involved in a chemical reaction?
     ✦ Reactants
295. Atoms of the same chemical element that have different atomic mass are known as?
     ✦ Isotopes
296. What is the fourth most abundant element in the universe in terms of mass?
     ✦ Carbon
297. What is the closest planet to the Sun?
     ✦ Mercury
298. What is the name of the 2nd biggest planet in our solar system?
     ✦ Saturn
299. What is the hottest planet in our solar system?
     ✦ Venus
300. What planet is famous for its big red spot on it?
     ✦ Jupiter
301. What planet is famous for the beautiful rings that surround it?
     ✦ Saturn
302. Can humans breathe normally in space as they can on Earth?
     ✦ No
303. Is the sun a star or a planet?
     ✦ A star
304. Who was the first person to walk on the moon?

✦ Neil Armstrong
305. What planet is known as the red planet?
✦ Mars
306. What is the name of the force holding us to the Earth?
✦ Gravity
307. Have human beings ever set foot on Mars?
✦ No
308. What is the name of a place that uses telescopes and other scientific equipment to research space and astronomy?
✦ An observatory
309. What is the name of NASA's most famous space telescope?
✦ Hubble Space Telescope
310. What is the name of the first satellite sent into space?
✦ Sputnik
311. Ganymede is a moon of which planet?
✦ Jupiter
312. What is the name of Saturn's largest moon?
✦ Titan
313. Olympus Mons is a large volcanic mountain on which planet?
✦ Mars
314. Is the planet Neptune bigger than Earth?
✦ Yes
315. Now that Pluto is no longer included, how many planets are there in the Solar System?
✦ 8
316. What is the smallest planet in the Solar System?
✦ Mercury
317. What is the largest planet in the Solar System?
✦ Jupiter
318. The chemical element uranium was named after what planet?
✦ Uranus
319. What planet in the solar system is farthest from the Sun?
✦ Neptune

Science Quiz

320. What is the second smallest planet in the solar system?
     + Mars
321. What planet is closest in size to Earth?
     + Venus
322. The moon Titan orbits what planet?
     + Saturn
323. The Galilean moons orbit what planet?
     + Jupiter
324. True or false? Venus has more atmospheric pressure than Earth?
     + True
325. Triton is the largest moon of what planet?
     + Neptune
326. What is the brightest planet in the night sky?
     + Venus
327. What is the third planet from the Sun?
     + Earth
328. Phobos and Deimos are moons of what planet?
     + Mars
329. What is the name of the biggest part of the human brain?
     + The cerebrum
330. The coloured part of the human eye that controls how much light passes through the pupil is called the?
     + Iris
331. What is the name of the substance that gives skin and hair its pigment?
     + Melanin
332. The muscles found in the front of your thighs are known as what?
     + Quadriceps
333. True or false? The two chambers at the bottom of your heart are called ventricles.
     + True
334. What substance are nails made of?
     + Keratin

335. What is the human body's biggest organ?
     - The skin
336. The innermost part of bones contains what?
     - Bone marrow
337. What is the number of bones in an adult human body?
     - 206
338. How many lungs does the human body have?
     - 2
339. Another name for your voice box is the?
     - Larynx
340. The two holes in your nose are called?
     - Nostrils
341. Your tongue is home to special structures that allow you to experience tastes such as sour, sweet, bitter, and salty, what is their name?
     - Taste buds
342. The bones that make up your spine are called what?
     - Vertebrae
343. The shape of DNA is known as?
     - A double helix
344. The flow of blood through your heart and around your body is called?
     - Circulation
345. The bones around your chest that protect organs such as the heart are called what?
     - Ribs
346. What is the name of the long pipe that shifts food from the back of your throat down to your stomach?
     - The oesophagus
347. True or false? Your ears are important when it comes to staying balanced.
     - True
348. The outside layer of skin on the human body is called the?
     - Epidermis
349. Name the longest rain forest in the World.

Science Quiz

- Amazon
350. Name the largest Mangrove forest in the World
    - The Sundarbans mangrove forest in West Bengal
351. Name the second largest Mangrove forest in the World
    - Pitchavaram in Tamil Nadu (near town of Chidambaram)
352. How many trees can be saved by recycling 1 ton paper?
    - 17 trees
353. When was the ozone layer hole over the Antarctic discovered?
    - 1985
354. How many litres of water is used to manufacture 1 ton of paper?
    - 55,000 litres.
355. Which city in India is least polluted?
    - Bangalore
356. Which country emits the largest amount of CO_2?
    - China
357. How long does it take for glass to decompose completely?
    - 1,000,000 years
358. Name the term used to denote substances that can be broken down by micro-organism.
    - Biodegradable
359. What % of World's oxygen content is emitted by Amazon rain forests?
    - 20%
360. Which day is called as "The International Day for the preservation of Ozone layer"?
    - September 16th.
361. Who was the first scientist to discover Green House effect?
    - Joseph Fourier (1824)
362. Which famous personality said "we won't have a society if we destroy the environment"?
    - Margaret Mead (American Anthropologist)
363. Where is the World's largest plastic recycling plant located?

- ✦ Beijing
364. Which is the berry used in the treatment of Ayurvedic and Unani medicine?
    - ✦ Gooseberry
365. In which year was the first world environmental day celebrated?
    - ✦ 1973
366. Which country is the world leader in the number of solar power systems installed per capita?
    - ✦ Kenya
367. The world's first commercial tidal power station was installed in 2007. Where was it?
    - ✦ Ireland
368. Environmental technology is also known as:
    - ✦ Envirotech
    - ✦ Greentech
    - ✦ Cleantech
    - ✦ **All of above**
369. Sea Levels are rising fast all around the globe. What is the highest level of rise over the past century?
    - ✦ 8 Inches
370. In which Country was Greenpeace Founded?
    - ✦ Canada
371. The Rio Treaty signed at 1992 Earth summit is officially known as
    - ✦ Convention on Biological Diversity
372. What is the name of the bi-monthly outreach journal of national tiger conservation authority published by Govt of India?
    - ✦ Stripes
373. Where is the office of the Forest Survey of India (FSI) located?
    - ✦ Dehradun
374. Every year the world gathers for the COP meetings – the "Conferences of the Parties" among the signatories of the

United Nations Framework Convention on Climate Change. When was this convention signed?
+ In 2002, in Johannesburg, as a follow-up to the Kyoto Protocol

375. At the 1992 Earth Summit two other major United Nations environmental conventions were signed in addition to the climate one, and they also hold their own COP meetings. Which are they?
+ The UN Convention on Biodiversity and the UN Convention to Combat Desertification

376. China is now the largest emitter of carbon dioxide (CO_2), followed by the United States. Who comes 3rd?
+ Japan

377. According to the Intergovernmental Panel on Climate Change (IPCC), what sectors are the largest emitters of greenhouse gases? (in order)
+ A. Transport sector, Energy generation, Agriculture

378. How much does the Question 9: The Kyoto Protocol requires developed countries to reduce their emissions compared to 1990 levels by 2020?
+ By 5.2% compared to 1990 levels.

379. One of the main arguments of developing countries is that, even if their total greenhouse gas emissions are increasing, their emissions per capita are still much lower than those of richer countries. Pick the correct statement:
+ The average US American emits 13x more than the average Indian

380. Which of these forms of biodiversity are considered particularly vulnerable to temperature changes in the global climate?
+ Small insects such as mosquitoes and flies

381. How high is the current rate of biodiversity loss compared to the natural rate?
+ 10x higher

382. Agricultural industrialization, too, has become seriously unsustainable. In the last 100 years, how much plant genetic diversity has been lost?
    ✦ 25%
383. The depletion of the ozone layer is not related to climate change, though many people mix up the two issues. What is the main cause of ozone depletion and of the hole in the ozone layer?
    ✦ CFCs, or chlorofluorocarbons, substances used in refrigerants and aerosols
384. Where is the hole in the ozone layer?
    ✦ Over the North Pole
385. When was the first time international leaders met with the purpose of discussing environmental issues?
    ✦ In 1992 at the United Nations Conference on Environment and Development
386. What do you think is the main cause of global climate change or the warming of the planet Earth?
    ✦ more carbon emissions
387. What percentage of the world's water is fresh and available for use?
    ✦ one percent (0.97%)
388. The current reduction in the number of certain species of salt-water fish is primarily due to?
    ✦ increased harvesting by fishing vessels
389. What is the leading cause of childhood death worldwide?
    ✦ germs in water
390. What is the most common reason that an animal becomes extinct?
    ✦ human destruction of habitats
391. There are thousands of waste disposal areas-dumps and landfills-in the U.S. that hold toxic waste. What is the greatest threat posed by the waste disposal areas (dumps and landfills)?
    ✦ contamination of water supplies

392. Some scientists have expressed concern that certain chemicals and minerals may accumulate in the human body at dangerous levels. Primarily, what do these chemicals and minerals enter the body through?
    + Drinking water
393. Earth's seasons are caused by which of the following?
    + The tilt of the earth's rotation relative to the ecliptic as earth revolves around the sun
394. Which of the following parts of the sun is easily visible only during a total solar eclipse?
    + Corona
395. What is most likely to cause a rise in the average temperature of earth's atmosphere in future?
    + CO_2 from fossil fuels
396. The accumulation of stress along the boundaries of lithospheric plates results in?
    + Earthquakes
397. Pollination by birds is called
    + Ornithophily
398. As you go down into a well, your weight
    + Decreases slightly
399. Which prefix is often used with scientific terms to indicate that something is the same, equal, or constant?
    + Iso
400. The study of phenomena at very low temperatures is called
    + Cryogenics
401. The branch of medical science which is concerned with the study of disease as it affects a community of people is called
    + Epidemiology
402. Superconductivity is a material property associated with
    + A loss of electrical resistance
403. If a metal can be drawn into wires relatively easily it is called
    + Ductile

404. Cystitis is the infection of which of the following?
    + urinary bladder
405. Water flows through a horizontal pipe at a constant volumetric rate. At a location where the cross-sectional area decreases, the velocity of the fluid
    + Increases
406. Yeast, used in making bread is a
    + Fungus
407. A gas used as a disinfectant in drinking water is
    + Chlorine
408. Vacuoles are bound by a definite membrane in plant cells called
    + Tonoplast
409. The theory which advocates that living beings can arise only from other living beings is termed
    + Bio-genesis
410. The wonder pigment chlorophyll is present in
    + Quantosomes
411. The solar eclipse occurs when
    + the moon comes in between the sun and the earth
412. The smallest functional and structural unit of kidney is called as
    + Nephron
413. The removal of top soil by water or wind is called
    + Soil erosion
414. The speed of light with the rise in the temperature of the medium
    + Remains unaltered
415. The oxide of Nitrogen used in medicine as anaesthetic is
    + Nitrogen pentoxide
416. The intensity of Earthquakes is measured on which scale?
    + Richter scale
417. The hardest substance available on earth is
    + Diamond
418. The gas predominantly responsible for global warming is

✦ Carbon dioxide
419. The dynamo is a device for converting
   ✦ Mechanical energy into electrical energy
420. The disease diphtheria affects which part of body?
   ✦ Throat
421. The concept of carbon credit originated from which Summit/Protocol?
   ✦ Kyoto Protocol
422. The cell that lacks a nucleus is
   ✦ Red blood corpuscles in man
423. The blue colour of the clear sky is due to
   ✦ Dispersion of light
424. The animal which uses sounds as its 'eyes' is
   ✦ Bat
425. Thalassaemia is a hereditary disease affecting
   ✦ Blood
426. Stainless steel is an example of
   ✦ A metallic compound
427. Small pox is caused by
   ✦ Virus
428. Silk is produced by
   ✦ Larva of silkworm
429. Persons with which blood group are called universal donors?
   ✦ O Negative
430. Persons with which blood group are called universal recipients?
   ✦ AB Positive
431. Plant cells can usually be distinguished because only plant cells possess
   ✦ Cell walls and central vacuoles
432. Who gave the first evidence of the Big-Bang theory?
   ✦ Edwin Hubble
433. Why is it difficult to see through fog?
   ✦ Rays of light are scattered by the fog droplets

434. Which types of waves are used in a night vision apparatus?
    + Infra-red waves
435. Which one of the following planets has largest number of natural satellites or moons?
    + Jupiter
436. A concrete wall generally,
    + absorbs and transmits sound
437. Which one of the following glands produces the Growth Hormone (Somatotropin)?
    + Adrenal
438. Which is a water-soluble vitamin?
    + Vitamin C
439. Which acid is produced when milk gets sour?
    + Lactic acid
440. When 1 litre of water freezes, the volume of ice formed will be
    + 1.11 litre
441. Water boils at a lower temperature on the hills because
    + There is a decrease in air pressure on the hills
442. The most abundant organic molecule on the surface of the Earth is
    + Cellulose
443. The persons working in textile factories such as carpet weavers are exposed to which occupational disease?
    + Asthma and Tuberculosis
444. The angle between the geographical meridian and magnetic meridian is called
    + Angle of declination
445. The characteristic odour of Garlic is due to?
    + Sulphur-containing compounds
446. Mist is caused by
    + Water vapours at low temperature
447. Earth quake waves travel fastest in
    + Water
448. Which colour of heat radiation represents the highest

temperature?
- ✦ White
449. What does airbag, used for safety of car driver, contain?
    - ✦ Sodium azide
450. The weight of an object will be minimum when it is placed at
    - ✦ The centre of the Earth
451. The uranium fuel used worldwide is mainly in the form of
    - ✦ UO2
452. Chlorination is a process used for water purification. The disinfecting action of chlorine is mainly due to
    - ✦ The formation of nascent oxygen when chlorine is added to water.
453. Chemically, bleaching powder is
    - ✦ Calcium hypochlorite
454. Deficiency of iron leads to
    - ✦ Anaemia
455. Radioactivity is measured by
    - ✦ Geiger-Muller counter
456. Regular intake of fresh fruits and vegetables is recommended in the diet since they are a good source of antioxidants. How do antioxidants help a person maintain health and promote longevity?
    - ✦ They neutralize the free radicals produced in the body during metabolism
457. The Himalayan Range is very rich in species diversity. Which one among the following is the most appropriate reason for this phenomenon?
    - ✦ It is a confluence of different bio-geographical zones
458. The function of heavy water in a nuclear reactor is to
    - ✦ Slow down the speed of neutrons
459. What is a zoom lens?
    - ✦ It is a lens having variable focal length
460. What is the difference between a CFL (Compact Fluorescent Lamp) and an LED (Light Emitting Diode) lamp?

- To produce light, a CFL uses mercury vapour and phosphor while an LED lamp uses semi-conductor material.
461. Seawater (i.e. saltwater) freezes at:
    - at a slightly lower temperature than fresh water.
462. Pollination by wind is called
    - Anemophily
463. The radioactive element most commonly detected in humans is
    - potassium-40
464. Radioisotopes which are used in medical diagnosis are known as
    - Tracers
465. Most commercial nuclear power plants worldwide are cooled by
    - Water
466. Which of the following materials is used along with iron ore and limestone to produce iron in a modern blast furnace?
    - Coke
467. The study of poisons is called
    - Toxicology
468. What range of frequencies are usually referred to as audio frequencies for humans?
    - 20 to 20,000 Hertz
469. A device used to measure the amount of moisture in the atmosphere is called a
    - Hygrometer
470. Plants get their nitrogen from
    - the soil
471. Steel is composed of a number of elements. The two essential elements are iron and
    - Carbon
472. Which of the following is an important element of stainless steel?
    - Chromium

Science Quiz

473. The most serious environmental pollution from a nuclear reactor is
     + thermal pollution
474. Biomass energy is
     + formed through photosynthesis
475. How much urine does the average person produce in a typical day?
     + 1.5 – 2 litres
476. Which country was the first to make fireworks?
     + China
477. Which sea is the saltiest natural lake and is also at the lowest elevation on the face of the earth?
     + The Dead Sea
478. Logarithms were devised by
     + John Napier
479. Name the process of large energy production in the Sun?
     + Nuclear fusion
480. Name the three common gases found in farts
     + CH_4 (methane); H_2S (hydrogen sulphide); NH_3 (ammonia)
481. Ornithology is the study of
     + Birds
482. Osteology is the study of
     + Bones
483. Palaeontology is the study of
     + Fossils
484. Phycology is the study of
     + Algae
485. Plant that eat insects are called
     + Insectivorous plants
486. The acid used in a car battery is
     + Sulphuric acid
487. The atomic number of oxygen is
     + Eight
488. What is the most common chemical element in the

universe?
- ✦ Hydrogen

489. What is the name for steel alloyed with chromium?
- ✦ Stainless steel

490. What is the name of the hormone that controls blood sugar level?
- ✦ Insulin

491. What is the name of the hot rock that is in the centre of the earth?
- ✦ Magma

492. What is the rest mass of a photon?
- ✦ Zero

493. What poisonous alkaloid is extracted from tobacco leaves and widely used as an insecticide (a chemical used for killing insects)?
- ✦ Nicotine

494. What yellow metal is an alloy of copper and zinc?
- ✦ Brass

495. Where does groundwater come from?
- ✦ Rainfall and melting snow

496. What do we call a person who cannot tell the difference between colours?
- ✦ Colour blind

497. What gas are the bubbles in Champagne?
- ✦ CO_2

498. What grain is beer usually made from?
- ✦ Barley

499. What is another name for the Palaeolithic Age?
- ✦ The stone age

500. What is E300?
- ✦ Vitamin C (Ascorbic acid)

501. What is the better-known name for the deadly poison prussic acid?
- ✦ Cyanide or hydrogen cyanide (HCN)

502. The symbol of silicon is

- Si
503. The symbol of silver is
    - Ag
504. The symbol Md stands for
    - Mendelevium
505. The symbol of sodium is
    - Na
506. The symbol of Sr stands for
    - Strontium
507. The symbol of titanium is
    - Ti
508. The symbol Rb stands for
    - Rubidium
509. The symbol Zn stands for
    - Zinc
510. The symbol Zr stands for
    - Zirconium
511. The two colours seen at the extreme ends of the pH chart are
    - Red & Blue
512. The unit of loudness is
    - Phon and
    - Sone
513. Thermostat is an instrument used for regulating
    - Constant temperature
514. Why food articles are mostly packed in aluminium foil?
    - To avoid rancidity
515. The basic building blocks of proteins are
    - Amino acids
516. The biggest plant seed is
    - Cocodemer (Lodoicea maldivica) (The Palm Family)
517. The botanical name for brinjal is
    - Solanum melongenal
518. The botanical name for onion is
    - Allium Cepa

519. The botanical name for rice is
    ✦ Oryza Sativa
520. The botanical name of tea is
    ✦ Camellia Sinensis
521. The botanical name of the cotton plant is
    ✦ Gossipium Hirsutum
522. The chemical formula of chloroform is
    ✦ $CHCl_3$ (Trichloromethane)
523. The chemical formula of common salt is
    ✦ NaCl
524. The chemical formula of lime soda is
    ✦ $CaCO_3$
525. The chemical formula of sodium bicarbonate is
    ✦ $NaHCO_3$
526. The formula H_2O_2 stands for
    ✦ Hydrogen peroxide
527. The formula H_2SO_4 stands for
    ✦ Sulphuric Acid
528. The formula HCl stands for
    ✦ Hydrochloric Acid
529. The liquid used to preserve specimens of plans and animals is
    ✦ Formalin
530. The metal used in storage batteries is
    ✦ Lead
531. The molecular formula of cane sugar is
    ✦ $C_{12}H_{22}O_{11}$
532. The response of a plant to heat is called
    ✦ Thermotropism
533. The response of a plant to touch is called
    ✦ Trigmotropism
534. The role of heredity was demonstrated by
    ✦ Mendel
535. The scientific name for blood platelets is
    ✦ Thrombocytes

536. The smallest flowering plant is
    - Wolffia (watermeal, or Wolffia globosa)
537. The study of antiquities is known as
    - Archaeology
538. The study of heavenly bodies is called
    - Astronomy
539. The study of sound is called
    - Acoustics
540. The study of tissues is called
    - Histology
541. Expand the following:
    - CFC — Chloro Fluoro Carbon
    - CCTV — Closed Circuit Television
    - AWACS — Airborne Warning and Control System
    - ATP — Adenosine TriPhosphate
    - AMU — Atomic Mass Unit
    - AYUSH — Ayurveda, Yoga & Naturopathy, Unani, Siddha and Homoeopathy
    - BCG — Bacillus Calmette Guerin
    - CDMA — Code Division Multiple Access
    - CNG — Compressed Natural Gas
    - CNS — Central Nervous System
    - COMPUTER — Commonly Operated Machine Particularly Used for Technology Education and Research
    - CRT — Cathode Ray Tube
    - DDT — Dichloro Diphenyl Trichloroethane
    - DNA — Deoxyribo Nucleic Acid
    - ECG — Electro Cardio Gram
    - FAX — Facsimile
    - FM — Frequency Modulator
    - GPRS — General Packet Radio Service
    - GPS — Global Positioning System
    - GSLV — Geostationary Launch Vehicle
    - GSM — Global System for Mobile Communications

- HAL — Hindustan Aeronautics Limited
- HIV — Human Immune deficiency Virus
- ICU — Intensive Care Unit
- ISCA — Indian Science Congress Association
- ISD — International Subscriber Dialing
- LED — Light Emitting Diode
- LASER — Light Amplification by Stimulated Emission of Radiation
- LCD — Liquid Crystal Display
- TFT — Thin Film Transistor
- LPG — Liquefied Petroleum Gas
- MASER — Microwave Amplification by Stimulated Emission of Radiation
- MCB — Miniature Circuit Breaker
- MMS — Multimedia Messaging Service
- MRI — Magnetic Resonance Imaging
- MP3 — MPEG Audio Layer 3
- NASA — National Aeronautic & Space Administration
- ODF — Open Defecation Free
- OPD — Out Patient Department
- PCO — Public Call Office
- PDA — Personal Digital Assistant
- pH — Potential of Hydrogen
- PNG — Portable Network Graphics
- PVC — Polyvinyl Chloride
- RADAR — Radio Detection and Ranging
- RNA — Ribose Nucleic Acid
- SAP — System Application & Products
- SARS — Severe Acute Respiratory Syndrome
- SIM — Subscriber Identity Module
- SMS — Short Message Service
- SONAR — Sound Navigation and Ranging
- STP — Standard Temperature and Pressure
- TB — Tuberculosis
- TNT — Tri Nitro Toluene

- ✦ UHF   Ultra High Frequency
- ✦ VHF   Very High Frequency
- ✦ VIRUS   Vital Information Resources Under Seize
- ✦ VLC   VideoLAN Client
- ✦ WiFi   Wireless Fidelity

542. Who was the first person to measure the speed of light?
    - ✦ Faucault
543. Who invented the air conditioner?
    - ✦ Carrier (Willis Haviland Carrier)
544. Who was the inventor of printing press?
    - ✦ Gutenberg
545. Who invented the microphone?
    - ✦ Berliner
546. Who invented the Polaroid camera?
    - ✦ Edwin Land
547. Who invented the aeroplane?
    - ✦ Wright Brothers (Orville & Wilbur Wright)
548. Who was the founder of the wave theory of light?
    - ✦ Christiaan Huygens
549. Who invented the commercial typewriter?
    - ✦ C. Sholes
550. Which scientist first determined the speed of light using a laboratory apparatus?
    - ✦ Louis Essen
551. Which hormone was the first to be discovered?
    - ✦ Secretin
552. Who invented Barometer?
    - ✦ Evangelista Torricelli
553. Who invented Ball-Point Pen?
    - ✦ John J. Loud
554. Who discovered DNA Structure?
    - ✦ Watson (US) and Crick (UK), Wilkins (UK)
555. Who invented Laser?
    - ✦ Charles H. Townes
556. Who invented Locomotive?

✦ George Stephenson
557. Who invented Logarithms?
✦ John Napier
558. Who discovered Circulation of blood?
✦ William Harvey
559. Who invented Vaccine of Rabies?
✦ Louis Pasteur
560. Who invented by Electric Lamp?
✦ Thomas Alva Edison
561. Who invented CT scan?
✦ Hounsfield
562. Television was invented by
✦ J. L. Baird
563. Electrons were discovered by
✦ J. J. Thompson
564. Roentgen discovered
✦ X–rays
565. The theory of Evolution was produced by
✦ Darwin
566. Neutron was discovered by
✦ Chadwick
567. Inert gases were discovered by
✦ Ramsay
568. Who discovered oxygen?
✦ Joseph Priestley
569. Thermo Flask was invented by
✦ Dewar
570. Streptomycin was invented by
✦ Walksman
571. Bushwell invented
✦ Submarine
572. Stethoscope was invented by
✦ Rane Laennec
573. Elias Home invented
✦ Sewing machine

Science Quiz

574. Pneumatic tyre was invented by
     ✦ Dunlop
575. The Solar System was discovered by
     ✦ Copernicus
576. Bicycle was invented by
     ✦ Macmillan
577. Cholera Bacillus was invented by
     ✦ Robert Koch
578. D.D.T. was invented by
     ✦ Paul Muller
579. Helicopter was invented by
     ✦ Broquet
580. Telegraphic code was introduced by
     ✦ Thomas Moore
581. Lift was invented by
     ✦ Otis
582. The cause of Beri Beri was discovered by
     ✦ Eijkman
583. Atomic theory was devised by
     ✦ John Dalton
584. Simpson and Harrison are associated with the invention of
     ✦ Chloroform
585. The field of activity of J.C. Bose was
     ✦ Botany
586. To an astronaut, the outer space appears
     ✦ Black
587. Fax machine was invented by
     ✦ Bain
588. Crescograph is an instrument to record
     ✦ plant growth
589. Lactometer is used to determine
     ✦ Purity of milk
590. Who is the father of discoveries?
     ✦ Archimedes
591. X–rays travel with the velocity of

✦ Light
592. Power of a lens is measured in
    ✦ Dioptre
593. Bhabha Atomic Research Centre (BARC) is located at
    ✦ Trombay
594. Power reactor is located in Tamil Nadu at
    ✦ Kalpakkam
595. What is used as a moderator in nuclear reactor
    ✦ Graphite
596. What is the unit of activity of a radioactive source?
    ✦ Becquerrel
597. Oil or soap film when in daylight appears coloured because of
    ✦ Interference
598. A woman's voice is shriller than man's voice due to
    ✦ higher frequency
599. Recording of sound on tapes was first invented by
    ✦ Poulsen
600. A watch based on the oscillating spring is taken from earth to moon, it will
    ✦ give the same time
601. When milk is churned cream gets separated due to
    ✦ centrifugal force
602. Earthquake is measured using
    ✦ Seismograph
603. The term Mach is used to measure speed of
    ✦ Aeroplane
604. The lens used to rectify long sight is
    ✦ convex lens
605. Einstein got the Nobel Prize for his theory of
    ✦ Relativity
606. An artificial satellite can be tracked very precisely from the earth by using
    ✦ Doppler effect
607. Solar cells are made up of

Science Quiz

- ✦ both silicon and germanium
608. The colour of star is an indication of its
   - ✦ Temperature
609. The hydraulic brakes used in automobiles is a direct application of
   - ✦ Pascal's law
610. A falling drop of rain water acquires the spherical shape due to
   - ✦ Surface tension
611. In a reactor, cadmium rods are used for
   - ✦ absorbing neutrons
612. Atom bomb is based on
   - ✦ nuclear fission
613. Raman effect involves
   - ✦ scattering of light
614. Ball pen works on the principles of
   - ✦ Capillarity and surface tension
615. Filament of an electric bulb is made of
   - ✦ Tungsten
616. The principle of Dynamo was discovered by
   - ✦ Michael Faraday
617. Magnetism at the centre of a bar magnet is
   - ✦ Zero
618. What do we call a substance which is repelled by a magnet?
   - ✦ Diamagnetic
619. A device used for measuring the depth of the sea is called
   - ✦ Fathometer
620. The source of sun's energy is
   - ✦ nuclear fusion
621. Land and sea breezes are due to
   - ✦ convection of heat
622. Which one of the following is also called Stranger Gas?
   - ✦ Xenon
623. The alcohol used in power alcohol is

- ethyl alcohol
624. Which synthetic fibre is known as artificial silk?
- Rayon
625. The chemical used as a fixer in photography is
- sodium thiosulphate
626. Nail polish remover contains
- Acetone
627. What is a mixture of potassium nitrate powdered charcoal and sulphur called?
- gun powder
628. Which is the heaviest metal?
- Mercury
629. Bleaching action of chlorine is by
- Decomposition
630. Bleaching powder contains
- Chlorine
631. pH of blood is
- 7.4
632. Bauxite is an ore of
- Aluminium
633. Natural rubber is a polymer derived from
- Isoprene
634. Which substance is obtained by the hydrolysis of oil?
- Glycerol
635. The alcohol used in the preparation of dynamite is
- Glycerol
636. The element used in lead pencils is
- Carbon
637. The solute in a solution can be separated by
- Evaporation
638. The concept of an electrolyte in water was described by
- Arrhenius
639. The term PVC used in the plastic industry stands for
- polyvinyl chloride
640. Normal valency of nitrogen is

Science Quiz

+ 3

641. The isotope atoms differ in
    + atomic weight
642. Which is used to produce artificial rain?
    + silver iodide
643. DDT is an
    + Insecticide
644. The concept of elliptical orbitals was suggested by
    + Sommerfeld
645. Yellow fever is spread by
    + house fly
646. Glucose is stored in our body in the form of
    + Glycogen
647. Rh factor is present in
    + Blood
648. The outer convex region of kidney is called
    + Cortex
649. Retina contains the sensitive cells called
    + rods and cones
650. Number of bones in human body is
    + 206
651. Birds excrete nitrogenous waste in the form of
    + uric acid
652. Night blindness is caused due to deficiency of
    + Vitamin A
653. In scorpion, poison is present in the
    + Sting
654. Rabies is transmitted by
    + infected mad dogs
655. The escape of haemoglobin from RBC is known as
    + Thermolysis
656. The branch of zoology which deals with the study of tissue is
    + Histology
657. The excretory organ of the earthworm is

- Nephridium
658. The rearing of silkworm is known as
    - Sericulture
659. Bile is secreted by
    - Liver
660. The enzyme which keeps the mouth clean from the action of bacteria is
    - Lysozyme
661. The blood vessels which carry blood towards the heart.
    - Veins
662. Kidneys are ------ shaped.
    - Bean
663. The tusks of an elephant are modified teeth called?
    - Incisors
664. Radioactive isotopes cause
    - Leukaemia
665. Saccharin is mainly used by
    - diabetic patients
666. Inner part of the brain is ----- in colour
    - White
667. How many pairs of spinal nerves are present in the human body
    - 31
668. Who is called Father of Genetics?
    - Mendel
669. Fixation of nitrogen in the soil
    - Bacteria
670. Who discovered nucleus?
    - Robert Brown
671. From which part of the plant is Clove obtained from
    - flower buds
672. Healing of wounds is hastened by vitamin
    - C
673. Pulses contain large amount of
    - Proteins

674. Another name of the reduction division is
     + Meiosis
675. Which plant was used by Mendel in his early experiment?
     + pea plant
676. Who found Binomial nomenclature?
     + Carl Linnaeus
677. More rate of photosynthesis occurs in
     + white light
678. The compound used in anti-malarial drug is
     + Chloroquine
679. Which one of the following vitamins is essential for bone and teeth formation?
     + Vitamin D
680. Food wrapped in newspaper is likely to get contaminated with
     + Lead
681. Azadiracta indica is the botanical name of
     + Neem
682. What is the main function of RNA?
     + protein synthesis
683. Paddy straw is stiff because of
     + Silica
684. Institute of Himalayan Bioresource Technology (Host Institution: CSIR - Council of Scientific and Industrial Research)
     + Palampur
685. Central Potato Research Institute (Host Institution: ICAR - Indian Council of Agricultural Research)
     + Shimla
686. Directorate of Mushroom Research (Host Institution: ICAR)
     + Solan Chambaghat
687. Himalayan Forest Research Institute
     + Shimla
688. GB Pant Institute of Himalayan Environment & Development

- Kullu
689. National Bureau of Plant Genetic Resource
    - Shimla
690. Institute of Integrated Himalayan Studies (IIHS)
    - Himachal Pradesh University, Shimla
691. Institute of Biotechnology & Environmental Science
    - Neri, Hamirpur
692. Indian Institute of Advance Study (IIAS)
    - Shimla
693. Himachal Pradesh Agriculture University
    - Palampur
694. North Temperate Regional Station (NTRS) of the Central Sheep and Wool Research Institute, Rajasthan (Host Institution: ICAR)
    - Garsa, Kullu
695. State Council for Science, Technology & Environment, Himachal Pradesh
    - Kasumpti, Shimla
696. Aryabhatta Geo-informatics & Space Application Centre (AGiSAC)
    - US Club, Shimla
697. State Centre on Climate Change, Himachal Pradesh
    - Shimla
698. Forest Research Institute (FRI)
    - Dehradun (Uttarakhand)
699. Tata Institute of Fundamental Research
    - Mumbai
700. National Dairy Research Institute
    - Karnal (Haryana)
701. National Aeronautical Laboratory
    - Bengaluru
702. All India Institute of Medical Sciences (AIIMS)
    - New Delhi
703. Indian Agricultural Research Institute
    - New Delhi

704. PGI Medical Education and Research
    + Chandigarh
705. Archaeological Survey of India
    + Kolkata
706. Central Scientific Instruments Organization
    + Chandigarh
707. Indian Institute of Science (IISc)
    + Bangalore, Karnataka
708. Indian Institute of Petroleum, (IIP), (CSIR)
    + Dehradun, Uttarakhand
709. Great Himalayan National Park
    + Kullu
710. Pin Valley National Park
    + Lahaul & Spiti
711. Wildlife Sanctuaries
    + Churdhar
        - Sirmour
    + Kalatop-Khajjiar
        - Chamba
    + Renuka
        - Sirmour
    + Simbalbara
        - Sirmour
    + Kibber (largest sanctuary by area in HP)
        - Lahaul & Spiti
    + Dhauladhar (2nd largest sanctuary by area in HP)
        - Kangra
712. What invention in about 1450 A.D. revolutionized communication and the world?
    + The Printing Press
713. What is the name for the new technology whereby a glass fibre carries as much information as hundreds of copper wires?
    + Fibre Optics (Opto-Electronics)
714. What mammal lays eggs?

- ✦ Platypus
715. How many time zones are there on Earth?
- ✦ Twenty-Four
716. What land mammal holds the record for the greatest age?
- ✦ Man
717. What colourless, pungent gas is often dissolved in water to yield a solution that is used as a biological preservative?
- ✦ Formaldehyde
718. What is the rest mass of a photon?
- ✦ Zero
719. What is the name given to the process, discovered by Goodyear, of adding sulphur to heated rubber?
- ✦ Vulcanization
720. Table sugar, from sugar cane or beet, is what type of sugar?
- ✦ Sucrose
721. Name the first woman to travel in space.
- ✦ Velentina Tereshkova
722. A type of plastic that is biodegradable has been in the news lately. The ingredient that makes it biodegradable is:
- ✦ Cornstarch
723. Who was the marine biologist and author of Silent Spring who was one of the first people to warn of the dangers of pesticides like DDT?
- ✦ Rachel Carson
724. What is the name given to the invasion of warm surface waters off the Peruvian coast that has been identified with strange climactic effects in recent years?
- ✦ El Nino
725. Occasionally, a bad cold will cause a decrease in a person's hearing ability. What is the name of the tube that becomes blocked to cause this problem?
- ✦ Eustachian Tube (pron: yu-sta-shen)
726. Name the general type of mammal that gives birth to undeveloped young that are kept in pouches.
- ✦ Marsupial

727. In which country was a method for making rust-resistant iron discovered in the fifth century B.C.?
    ✦ India
728. At room temperature, most elements are in which phase of matter?
    ✦ Solid
729. What compound is a common component of air pollution, but is essential in the upper atmosphere to protect life on earth?
    ✦ Ozone
730. What radioactive element is routinely used in treating hyperthyroidism, and in reducing thyroid activity?
    ✦ Iodine-131
731. What is osteoporosis?
    ✦ Loss of Calcium from Bones
732. The Statue of Liberty is green because of:
    ✦ Oxidized Copper
733. At what point are the Celsius and Fahrenheit scales equal?
    ✦ - 40 Degrees
734. The word atom is from a Greek word meaning:
    ✦ Indivisible
735. Identify the Earth's largest INVERTEBRATE animal.
    ✦ Giant Squid
736. What creature do the Galapagos Islands take their name from?
    ✦ Tortoise
737. What is shed when you desquamate?
    ✦ Skin or Epidermal cells
738. Nitrous oxide, commonly called laughing gas, has been a matter of concern to environmentalists recently because:
    ✦ It is a Greenhouse gas.
739. Ozone in the upper atmosphere is produced from:
    ✦ Photochemical Reactions
740. What is the technique by which lasers are used to photograph objects and reproduce them in three

dimensions?
- ✦ Holography

741. Name the effect that is caused by scattering of light in colloids or suspensions.
    - ✦ Tyndall Effect

742. Name the scientist who won the Nobel Prize in 1962 for elucidating the details of the photosynthetic process.
    - ✦ (Melvin) Calvin

743. Atolls are most probably formed by:
    - ✦ Coral Reef Upbuilding During Subsidence

744. The earliest experiments on chromosome theory and inheritance were developed by?
    - ✦ Gregor Mendel

745. The art of growing dwarfed trees in small pots, a technique perfected by the Japanese, is known as what?
    - ✦ Bonsai

746. When natural uranium is mined, it contains three isotopes. Which TWO are important in the production of nuclear power?
    - ✦ Uranium-235 And Uranium-238

747. In a high-temperature gas-cooled reactor, what gas may be used for a coolant?
    - ✦ Helium or Carbon Dioxide

748. Spent fuel from a nuclear power plant cools down and loses most of its radioactivity through decay. In the first year, spent fuel loses about what percentage of its radioactivity?
    - ✦ 80 Percent

749. Which country generates the greatest percentage of its electricity by nuclear power?
    - ✦ France

750. What substance was used as a moderator for the chain reaction in the first nuclear reactor?
    - ✦ Graphite

751. What do you produce by adding Lactobacillus Bulgaricus to

milk?
+ Yogurt

752. What month is showing on the calendar when the Earth is nearest the sun?
+ January

753. On June 22, how many hours of daylight are there in a day at the North Pole?
+ 24

754. How many minutes does it take for light to travel from the sun to the Earth?
+ 9-1/2 Minutes

755. What is the principal liquid used in automotive antifreeze?
+ Ethylene Glycol

756. What are the names of the cycles in a 4-stroke gasoline engine?
+ Compression; Ignition; Power; Exhaust

757. If an individual is suffering from emphysema, what organ of the body is primarily affected?
+ Lungs

758. At what Celsius temperature is water the densest?
+ 4 Degrees Celsius

759. What household heating fuel also powers jet planes?
+ Kerosene

760. Who discovered the method of destroying disease-producing bacteria in beer and milk by heating the liquid properly?
+ (Louis) Pasteur

761. What is the name denoting plants that seasonally shed all their leaves?
+ Deciduous

762. What is meant by the statement that an animal is oviparous?
+ It Lays Eggs

763. What is the distinction shared by Marie Curie and Linus Pauling?

- They Both Received Two Nobel Prizes
764. What is the mass in grams of one cubic centimetre of water?
    - One Gram
765. You use a hydrometer to check the condition of the electrolyte in your car battery. It is actually measuring what property of the fluid?
    - Density
766. The principle behind a breeder reactor is:
    - U-238 Is Converted to Pu-239, Which Undergoes Fission.
767. When petroleum is refined it is subject to a number of processes including cracking. During the cracking process
    - Heavy hydrocarbons are broken up into lighter ones
768. One metal dissolved in another is called
    - An Alloy
769. The most suitable motor oil for use in a car in cold weather is:
    - A Low Viscosity Oil
770. The process by which a substance absorbs moisture upon exposure to the atmosphere is called:
    - Deliquescence
771. The first animal launched into orbit was a:
    - Dog
772. Name the first man-made space craft to orbit the Earth?
    - Sputnik
773. "Fool's gold" is a common name for this mineral:
    - Pyrite
774. Which two gases are used to disinfect water in sewage treatment facilities?
    - Ozone and Chlorine
775. What naturally occurring radioactive substance present in indoor air is suspected as the second leading cause of lung cancer?
    - Radon

776. Thermal insulation is used to:
     + Reduce the flow of heat
777. The oestrogen and androgen hormones which regulate sexual development and function are structurally related to:
     + Cholesterol
778. The most abundant organic molecule on the surface of the Earth is:
     + Cellulose
779. Liquid fuels can be used to generate electricity in a variety of ways. Which of the following technologies converts liquid fuel directly to electricity without a combustion step?
     + Fuel Cell
780. What is the approximate weight in pounds of the atmosphere on a 1 sq. ft. surface at sea level?
     + 2100 LBS.
781. What common flavouring comes from the long slender fruit of a climbing orchid?
     + Vanilla
782. In daylight the human eye is most sensitive to which colour?
     + Green
783. What is the name for the metric unit of pressure defined as 1 Newton per meter squared?
     + Pascal
784. Who was the first scientist credited with pointing out that certain gases could cause a greenhouse effect?
     + John Tyndall
785. The sun makes up approximately what percent of matter in our solar system?
     + 99%
786. Movement exhibited by plants toward the "centre of the Earth" is called?
     + Geotropism
787. Who is known as the Father of the Atomic Bomb?

- J. Robert Oppenheimer
788. Evaporation from water surfaces exposed to air is not dependent of the:
    - Depth of the Water
789. What is the SI unit of length?
    - Metre (m)
790. What is a light year?
    - The distance travelled by light in one year.
    - 1 light year = 9.46×10^{15} m
791. What is Astronomical year?
    - The mean distance from the centre of earth to the centre of the sun.
    - 1 A.U. = 1.495×10^{16} m
792. What is the SI unit of mass?
    - Kilogram (kg)
793. What is the SI unit of time?
    - Second
794. What acts like a pendulum in atomic clock?
    - Caesium atoms (atomic clock works on energy changes in gaseous caesium atoms)
795. What is the SI unit of speed?
    - Metre per second
796. What is the SI unit of velocity?
    - Metre per second
797. Speed is a:
    - Scalar quantity
798. Velocity is a:
    - Vector quantity
799. What is acceleration?
    - It is rate of change of velocity
800. What is the value of the acceleration due to gravity (g) on the surface of earth?
    - 9.8 m/s^2 (metre per second square)
801. "To every action there is an equal and opposite reaction".
    - Newton's Third Law

Science Quiz

802. What is the SI unit of work?
    + Joule (J)
803. What is the SI unit of power?
    + Watt (W)
804. What is the SI unit of pressure?
    + Pascal
805. Valency of first 20 elements:

Atomic No.	Symbol	Element	Valency (most common)
1	H	Hydrogen	+1
2	He	Helium	0
3	Li	Lithium	+1
4	Be	Beryllium	+2
5	B	Boron	+3
6	C	Carbon	+4, -4
7	N	Nitrogen	+5, -3
8	O	Oxygen	-2
9	F	Fluorine	-1
10	Ne	Neon	0
11	Na	Sodium	+1
12	Mg	Magnesium	+2
13	Al	Aluminium	+3
14	Si	Silicon	+4
15	P	Phosphorus	+5, -3
16	S	Sulphur	+6, -2
17	Cl	Chlorine	-1
18	Ar	Argon	0
19	K	Potassium	+1
20	Ca	Calcium	+2

806. What is the SI unit of pressure?
    + Pascal
807. Kaziranga National Park
    + Assam

808. Manas National Park
    + Assam
809. Gir Forest National Park
    + Gujarat
810. Kalesar National Park
    + Haryana
811. Inderkilla National Park
    + Himachal Pradesh (Kullu)
812. Khirganga National Park
    + Himachal Pradesh (Kullu)
813. Simbalbara National Park
    + Himachal Pradesh (Sirmaur)
814. Jim Corbett National Park
    + Uttarakhand
815. The Nobel Prize in Physics 2016
    + for theoretical discoveries of topological phase transitions and topological phases of matter
        - David J. Thouless
        - F. Duncan M. Haldane
        - J. Michael Kosterlitz
816. The Nobel Prize in Chemistry 2016
    + for the design and synthesis of molecular machines
        - Jean-Pierre Sauvage
        - Sir J. Fraser Stoddart
        - Bernard L. Feringa
817. The Nobel Prize in Physiology or Medicine 2016
    + for his discoveries of mechanisms for autophagy
        - Yoshinori Ohsumi
818. Chandra Shekar Venkata Raman/Sir C.V. Raman-1930
    + Sir. C.V. Raman, an Indian Scientist / Physicist was awarded Nobel Prize of Physics in 1930 for his "Raman Effect" related to light.
819. Dr. Hargobind Khorana-1968
    + Dr. Hargobind Khorana, an India's Doctorate in Chemistry was awarded Nobel Prize for Medicine in

1968 for his study of the Human Genetic Code and its role in Protein Synthesis.

820. Dr. Subramanian Chandrashekar-1983
   - Dr. Subramanian Chandrashekar, an Indian Astro-Physicist was awarded Nobel Prize for Physics in 1983 for his theory on white dwarf stars' limitation known as 'Chandrasekhar Limit'. Dr. Subramaniyan is the son of the elder brother of the 'Nobel Prize Winner'- Sir C.V. Raman.

821. Dr. Venkataraman Ramakrishnan-2009   (Foreign citizen of Indian origin)
   - Dr. Venkataraman Ramakrishnan, an Indo-American has shared Nobel Prize for Chemistry for the year 2009, along with a co-American Thomas Steitz and Ada Yonath of Israel for mapping ribosomes, the protein producing factories within cells at the atomic level.

822. Other Nobel Prize Winners with Indian connections:
   - Ronald Ross: Nobel Prize in Physiology/Medicine (1902) Foreign citizen born in India

823. Indian Scientists awarded Bharat Ratna:
   - Prof C N R Rao
   - Sir C.V. Raman
   - Former president Dr. A P J Abdul Kalam

824. Famous Indian Scientists and their Inventions
   - Prafulla Chandra Ray
     - Famous academician and chemist, known for being the founder of Bengal Chemicals & Pharmaceuticals, India's first pharmaceutical company.
   - Salim Ali
     - Naturalist who helped develop Ornithology; also known as the "birdman of India".
   - Srinivasa Ramanujan
     - Mathematician known for his brilliant contributions to contributions to mathematical

analysis, number theory, infinite series, and continued fractions.

- **C. V. Raman**
  - Physicist who won Nobel Prize in 1930 for his Raman Effect.
- **Homi Jehangir Bhabha**
  - Theoretical physicist; best known as the chief architect of the Indian atomic energy program.
- **Jagadish Chandra Bose**
  - Physicist, biologist, and archaeologist who pioneered the investigation of radio and microwave optics.
- **Satyendra Nath Bose**
  - Mathematician and physicist; best known for his collaboration with Albert Einstein in formulating a theory related to the gas like qualities of electromagnetic radiation.
- **A.P.J. Abdul Kalam**
  - Known for his crucial role in the development of India's missile and nuclear weapons programs.
- **Har Gobind Khorana**
  - Biochemist who won the Nobel Prize in 1968 for demonstrating how the nucleotides in nucleic acids control the synthesis of proteins.
- **S.S. Abhyankar**
  - Mathematician; famous for his outstanding contributions to algebraic geometry.
- **Meghnad Saha**
  - Astrophysicist who developed the Saha equation, which explains chemical and physical conditions in stars.
- **Subrahmanyan Chandrasekhar**
  - Astrophysicist won the Nobel Prize in 1983 for his research on the evolutionary stages of massive stars.

Science Quiz

- Raj Reddy
  - A.M. Turing Award-winning computer scientist, best known for his work related to large scale artificial intelligence systems.
- Birbal Sahni
  - Palaeobotanist known for his research on the fossils of the Indian subcontinent.
- Prasanta Chandra Mahalanobis
  - Statistician and physicist who founded the Indian Statistical Institute.
- Prof C N R Rao:
  - Indian chemist who has worked mainly in solid-state and structural chemistry.

825. Iodine is used to test for which substance in food?
    - Starch
826. What is the scientific name for the kneecap?
    - Patella
827. Which is the most acidic part of the digestive system?
    - Stomach
828. What is the term for a positive electrode?
    - Anode
829. What is the lightest metal?
    - Lithium
830. Where in the human body do you find the alveoli?
    - Lungs
831. A bone is joined to a muscle by what structure?
    - Tendon
832. What device is used to vary AC voltage?
    - Transformer
833. Which gas smells like rotten eggs?
    - Hydrogen Sulphide
834. What is the name of the device used to convert sound to electricity?
    - Microphone
835. What is protected by the cranium?

+ Brain

836. What tube connects the kidney to the bladder?
    + Ureter
837. Where are the red blood cells made?
    + Bone Marrow
838. How many chromosomes are in the nucleus of a human sperm?
    + 23
839. What do the letters pH stand for?
    + Power of Hydrogen
840. How many teeth should an adult have including their wisdom teeth?
    + Thirty-two (32)
841. Which two parts of the body continue to grow for your entire life?
    + nose and ears
842. What makes up 80% of our brain?
    + Water
843. In the adult human body, what is longer - the small intestine or the large intestine?
    + The small intestine
844. Where is the smallest bone in your body?
    + Ears
845. Where is the largest bone in your body?
    + Leg
846. How many ribs are there in the human body?
    + Twenty-four (24). Humans have 12 pairs of ribs / 24 ribs in total
847. What is the strongest muscle in your body?
    + Your tongue
848. Which colour eyes do more humans have?
    + Brown
849. Who created the famous equation: $E = mc^2$
    + Albert Einstein
850. In which part of the human body can the longest muscle be

found?
- ✦ Thigh (leg)

851. Dry ice is a frozen form of which gas?
- ✦ Carbon Dioxide

852. What term is given to a piece of rock or metal from space that reaches the surface of the Earth?
- ✦ Meteorite

853. Who invented the jet engine?
- ✦ Sir Frank Whittle

854. What would you use VOIP for?
- ✦ Making a telephone call on the internet (it stands for Voice Over Internet Protocol)

855. Which subatomic particles are found in the nucleus of an atom?
- ✦ Protons and Neutrons

856. Orbiting 35,900km above the equator, what term is given to satellites that remain above the same point on the Earth's surface in their orbit?
- ✦ Geostationary

857. Which chemical element, number 11 in the Periodic table, has the symbol Na?
- ✦ Sodium

858. What is the largest fish in the world?
- ✦ The whale shark

859. What is the longest bone in the human body?
- ✦ The femur (or thighbone)

860. Relating to flat-screen televisions and monitors, what does LCD stand for?
- ✦ Liquid Crystal Display

861. Diamonds are a form of which chemical element?
- ✦ Carbon

862. What is the device that blends air and fuel for an internal combustion engine called?
- ✦ Carburettor

863. Magnetite, hematite, limonite, and siderite are ores of

which metal?
+ Iron

864. What name is given to the condition created by too much bile in the bloodstream creating a yellowing of the skin?
+ Jaundice

865. What acid accumulates in the muscles once the anaerobic threshold is passed when doing exercise?
+ Lactic Acid

866. What do 1,000 gigabytes make?
+ Terabyte

867. Which sugar is found in milk?
+ Lactose

868. What sort of structure is DNA?
+ double helix

869. In which spacecraft did Yuri Gagarin become the first man in space?
+ Vostok 1

870. Riboflavin is an alternative name for which vitamin of the B Group?
+ Vitamin B2

871. Which instrument is used to measure pressure?
+ Manometer

872. The substance that triggers the fall of mature leaves and fruits from plants is
+ Abscisic acid

873. Who is the president of the Council of Scientific and Industrial Research?
+ Prime Minister of India

874. Which was the India's first operational remote sensing satellite?
+ IRS-1A

875. Name India's first satellite exclusively used for educational purpose?
+ EDUSAT

876. From where was India's first mission to Moon

Science Quiz

Chandrayaan-1 launched?
- Sriharikota

877. Which was India's first indigenous satellite launch vehicle?
- SLV-3

878. Who was the first chairperson of Indian Space Research Organisation (ISRO)?
- Vikram Sarabhai

879. Who is known as the father of Indian nuclear programme?
- Homi Jehangir Bhabha

880. What was the code name of India's first nuclear test in India in the year 1974?
- Smiling Buddha

881. Where is the headquarter of National Academy of Sciences located?
- Allahabad

882. From where was India's first satellite Aryabhata launched?
- Soviet Union

883. First satellite to be placed in orbit by Indian made launch vehicle SLV-3
- Rohini

884. When was India's first satellite Aryabhatta launched by India?
- 1975

885. Who is known as the father of Indian space program
- Vikram Sarabhai

886. First mobile phone conversation took place in India between?
- Jyoti Basu and Sukhram

887. Internet in India was started by VSNL in the year?
- 15th August 1995

888. First nuclear plant in India
- Tarapur, Maharashtra

889. First nuclear reactor made in India
- Apsara

890. Where was the first hydro-electric power plant in India?

- Darjeeling
891. Where is ISRO headquarter located?
    - Bangalore
892. India's first supercomputer is known as:
    - PARAM 8000
893. Who is known as the architect of PARAM series supercomputers?
    - Vijay P. Bhatkar
894. Who is the creator of world's second and India's first test tube baby Durga?
    - Subhash Mukhopadhyay
895. Name the first IIT established?
    - IIT Kharagpur
896. Bhakra dam is built on which river?
    - Sutlej
897. Which day is celebrated as National Mathematics Day in India as a respect to S Ramanujan?
    - 22 December
898. S Ramanujan is the first Indian to be elected as fellow of which college?
    - Trinity College
899. Which number if famously known as Ramanujan-Hardy number?
    - 1729
900. Where is the headquarter of Indian Science Congress Association located?
    - Kolkata
901. Who was the president of the first Indian Science Congress meeting held in Kolkata in 1914?
    - Ashutosh Mukherjee
902. Which is the biggest planetarium in India?
    - Birla Planetarium
903. Who is the founder of Indian Institute of Science, Bangalore?
    - Jamsetji Tata

904. When the first expedition to Antarctica was started by India?
    ✦ 1981
905. Who was the leader of the first expedition to Antarctica started by India?
    ✦ Sayed Z. Qasim
906. What is India's permanent research station in Antarctica?
    ✦ Maitri
907. What was India's first permanent base station in Antarctica?
    ✦ Dakshin Gangotri
908. Where is the headquarter of Indian Statistical Institute located?
    ✦ Kolkata
909. In which year was Indian Statistical Institute established?
    ✦ 1931
910. Who is known as the father of electricity?
    ✦ Benjamin Franklin
911. Who invented telescope?
    ✦ Hans Lipperhey
912. When the Nobel prizes were first awarded?
    ✦ 1901
913. Who is the first woman to win a Nobel prize?
    ✦ Marie Curie
914. Nobel Prize except the Peace Prize in awarded at which city?
    ✦ Stockholm
915. Who is known as the father of evolution?
    ✦ Charles Darwin
916. Who is known as the inventor of hydrogen bomb?
    ✦ Edward Teller
917. What was the code name of the first nuclear test conducted by USA on 16th July 1945?
    ✦ Trinity
918. Who first developed the polio vaccine?

+ Jonas Salk
919. Edward Jenner is famous for the invention of
+ Smallpox vaccine
920. Which gas is also known as laughing gas?
+ Nitrous oxide
921. Who is known as the inventor of steam engine?
+ James Watt
922. Which vitamin is produced by sunlight?
+ Vitamin D
923. Who is known as the father of biology?
+ Aristotle
924. Who first distinguished the main blood groups in human body?
+ Karl Landsteiner
925. Which is the closest star to the solar system?
+ Proxima Centauri
926. Halley's Comet last appeared in which year?
+ 1986
927. Who is the first female commander of a space shuttle?
+ Eileen Collins
928. Where is the headquarter of NASA located?
+ Washington
929. Which gas is used for flying balloons?
+ Helium
930. Which is the nearest planet to the earth?
+ Venus
931. The disease scurvy is caused by lack of
+ Vitamin C
932. Which is the first surface to surface missile in India?
+ Prithvi
933. Total number of chromosomes per cell in human body?
+ 46
934. The vitamin that helps in the clotting of blood is
+ Vitamin K
935. Who discovered cell (building blocks of life)?

Science Quiz

- ✦ Robert Hooke
936. Total how many SI units are defined by International System of Units (SI)?
    - ✦ 7
937. Total number of bones in a human spine
    - ✦ 33
938. Who performed the world's first successful human-to-human transplant?
    - ✦ Christian Barnard
939. What was the name of the first space shuttle on which Kalpana Chawla flew to space in 1997?
    - ✦ Columbia
940. In which year the disaster of space shuttle Columbia happened in which Kalpana Chawla died?
    - ✦ 2003
941. Which is the only movable bone in the human skull?
    - ✦ Mandible
942. In which year did Yuri Gagarin become the first person in space?
    - ✦ 1961
943. Total number of bones in human skull
    - ✦ 22
944. Which cells are responsible to send and receive electro-chemical signals to and from the brain?
    - ✦ Neurons
945. Once Celsius is equal to how many Fahrenheit?
    - ✦ 32
946. In the periodic table, how many elements are found naturally
    - ✦ 98
947. What is the total number of elements currently in the periodic table?
    - ✦ 118
948. When was the periodic table first published by Dmitri Mendeleev?

- 1869
949. Which is the first element in the periodic table?
    - Hydrogen
950. Aedes aegypti mosquito bite can cause
    - Dengue
951. The outermost layer of sun is called
    - Corona
952. What is the speed of light?
    - 299,792,458 m/s
953. What is the speed of sound?
    - 340.29 m/s
954. Which gas is usually filled in the electric bulb?
    - Nitrogen
955. Which element has the highest electronegativity?
    - Fluorine
956. From which types of solar radiation does sunscreen protect the skin from?
    - Ultraviolet
957. Glass is made of the mixture of
    - Sand and silicates
958. Kerosene is a mixture of
    - Hydrocarbons
959. What is the general pH level of human skin?
    - 5.5
960. Deuterium oxide is also known as
    - Heavy water
961. The blue colour of the clear sky is due to
    - Dispersion of light
962. Which planet is also known as morning star and evening star
    - Venus
963. The list of endangered species is released by
    - IUCN (International Union for Conservation of Nature and Natural Resources)
964. The colour of tomatoes is due to the presence of

✦ Flavanoids
965. Which part of the body is affected by the trachoma disease?
✦ Lungs
966. Radiowaves are reflected back to earth from which layer?
✦ Ionosphere
967. Which fuel is used in gas welding?
✦ Acetylene
968. Lipids are
✦ Fats of natural origin
969. If a ship moves from freshwater to seawater, it will
✦ Raise a little higher
970. Polio is caused by the infection of
✦ RNA virus
971. What is measured by anemometer?
✦ Speed of wind
972. What is measured by Ohmmeter?
✦ Electrical resistance
973. Biogas is a mixture of
✦ Methane and carbon dioxide
974. The average life span of Red Blood Cell is
✦ 120 days
975. Cooking gas is a mixture of
✦ Butane and propane
976. What is the primary test of diagnosing HIV and AIDS?
✦ ELISA Test (Enzyme-linked Immunosorbent Assay
977. What is the confirmation test for HIV and AIDS?
✦ Western Blot Test
978. The most suitable unit for expressing nuclear radius is
✦ Angstrom
979. Which scientist proved that the path of each planet around the Sun is elliptical?
✦ Kepler
980. The king of metals is
✦ Gold

981. Who discovered the link between electricity and magnetism?
    + Michael Faraday
982. Glaucoma affects which parts of the body?
    + Eyes
983. Entomology is the science that studies
    + Insects
984. Saliva helps in the digestion of
    + Starch
985. Sea breeze is formed during
    + Day time
986. Diabetes is caused by dysfunctioning of
    + Pancreas
987. Which gas is used for artificial ripening of green fruits?
    + Acetylene
988. The hormone insulin is a
    + Glycolipid
989. In which organ of the human body, lymphocyte cells are formed?
    + Long bone
990. The pH value of blood is
    + 7.4
991. Cerebral malaria is caused by
    + Plasmodium falciparum
992. Name the latest communication satellite launched by India:
    + India's latest communication satellite GSAT-18 was successfully launched by a heavy-duty rocket of Arianespace from the spaceport of Kourou in French Guiana in October 2016
993. Recent PSLV launch:
    + The PSLV C35 Polar Satellite Launch Vehicle, successfully launches 8 satellites in September 2016
994. Name India's fastest and most powerful computer
    + PARAM-ISHAN
995. Where is Param-Ishan supercomputing facility located?

- IIT, Guwahati
996. Who launched Param-Ishan
    - Union Human Resource Development Minister Prakash Javadeka in September 2016
997. India launched INSAT-3DR, an advanced weather satellite in September 2016 from Sriharikota in Andhra Pradesh through vehicle
    - GSLV-F05 which is the tenth flight of India's Geosynchronous Satellite Launch Vehicle (GSLV)
998. Indian Space Research Organisation (ISRO) successfully tested its own scramjet engines from Satish Dhawan Space Centre (SDSC) in Sriharikota in Andhra Pradesh
    - In August 2016
999. In January 2014, ISRO successfully used an indigenous cryogenic engine in a GSLV-D5 launch of the GSAT-14.
1000. ISRO sent one lunar orbiter, Chandrayaan-1, on 22 October 2008 and one Mars orbiter, Mars Orbiter Mission, which successfully entered Mars orbit on 24 September 2014, making India the first nation to succeed on its first attempt, and ISRO the fourth space agency in the world as well as the first space agency in Asia to successfully reach Mars orbit.
1001. Polar Satellite Launch Vehicle (PSLV) C-34
    - ISRO successfully launched record 20 satellites in a single mission in June 2016
1002. RLV-TD
    - Indigenous technology demonstrator of reusable launch vehicle tested successfully in May 2016. Capable of launching satellites into orbit around earth and re-entering the atmosphere
1003. ISRO Scientists made world's lightest material (synthetic made by man)
    - 'Silica Aerogel' or blue air (in April 2016)
1004. India's first open source encyclopaedia to Art and Culture
    - Sahapedia

Science Quiz

1005. Fastest supercomputer in the world
    ✦ China's supercomputer Tianhe-2 is the fastest in the world
1006. Leakage of which gas had caused the Bhopal Gas Tragedy?
    ✦ Methyl isocyanate
1007. One carbon credit is equivalent to –
    ✦ 1000 kg of CO2 (A carbon credit is a generic term for any tradable certificate or permit representing the right to emit one tonne of carbon dioxide or the mass of another greenhouse gas with a carbon dioxide equivalent (tCO2e) equivalent to one tonne of carbon dioxide.)
1008. Which gas is most responsible for global warming?
    ✦ Carbon di-oxide
1009. Who invented Telegraph?
    ✦ Samuel Morse
1010. The apparatus used for detecting lie is known as –
    ✦ Polygraph
1011. Cryogenic engines find applications in –
    ✦ Rocket technology
1012. Communication satellite are used to –
    ✦ receive and redirect communication signal
1013. Who discovered the fact that - "An electric current produces a circular magnetic field as it flows through a wire"?
    ✦ Oersted
1014. A radar that detects the presence of an enemy aircraft uses –
    ✦ Radio waves
1015. Vermiculture technology is used in:
    ✦ Organic Farming
1016. Date of manufacture of food items fried in oil should be checked before buying because oils become rancid due to –
    ✦ Oxidation
1017. Which of the following instruments can be used for

measuring the speed of an aeroplane –
+ Pilot tube

1018. 'White Revolution' is related to
+ Milk Production

1019. The basic function of technology 'Blue Tooth' is to allow
+ wireless communication between equipment

1020. Name the country which launched the first Satellite "Sputnik" into the space.
+ Soviet Union

1021. Human Environment Conference 1972 was held at
+ Stockholm

1022. Who invented the video-tape?
+ Charles Ginsberg

1023. The 'Param' series of supercomputers was developed in India by which of the following institutions?
+ Centre for Development of Advance Computing (CDAC)

1024. Who was the founder Director of the TIFR (Tata Institute of Fundamental Research)
+ Homi Jehangir Bhabha

1025. When was colour TV transmission introduced in India?
+ 1982

1026. Who is recognised as the Father of Geometry?
+ Euclid

1027. How many countries have exploded the atom bomb before India?
+ 5 (India's first atomic bomb was detonated on May 18, 1974. Before that Soviet Union, USA, Britain, France, and China had exploded the atom bomb.)

1028. Fountain pen was invented by
+ Lewis E. Waterman

1029. What type of electromagnetic radiation is used in the remote control of a television?
+ Infrared

1030. Refrigeration helps in food preservation by –
+ Reducing the rate of biochemical reaction

1031. Zoological Survey of India has it headquarter at –
    ✦ Kolkata
1032. Carbon dioxide is called greenhouse gas because-
    ✦ it absorbs infrared radiation
1033. Name the process of production of energy in the Sun
    ✦ Nuclear fusion
1034. Who invented chloroform as anaesthetic?
    ✦ James Simpson
1035. Which international agreement was ratified by India on 2nd October 2016?
    ✦ India ratified the Paris Agreement (on Climate Change)
1036. Electrostatic precipitator is used to control
    ✦ Air pollution
1037. Penicillin, an, antibiotic, is obtained from a
    ✦ Fungus
1038. Which elements (metals) pollutes the air of a city having large number of automobiles?
    ✦ Cadmium
1039. Where was India's first computer installed?
    ✦ Indian Statistical Institute, Calcutta
1040. Hydrogen bomb is based on the principle of
    ✦ uncontrolled fusion reaction
1041. Who invented penicillin?
    ✦ Alexander Fleming
1042. The name of the white revolution is associated with
    ✦ Verghese Kurien
1043. Which branch deals with the interactions of same species of living organisms with their non-living environment?
    ✦ Ecology
1044. The scientist associated with the success of Green Revolution is
    ✦ Norman Borlaug
1045. Norman Ernest Borlaug, who is regarded as the father of the Green revolution in India, is from which country?
    ✦ USA

Science Quiz

1046. World Environment Day is Celebrated Every Year on
    + June 5
1047. What is RDX?
    + An explosive
1048. Centre for DNA Fingerprinting is located at
    + Hyderabad
1049. The device used for locating submerged object sunder sea is
    + Sonar
1050. The branch of study dealing with old age and aging is called-
    + Gerentology
1051. Which famous Indian scientist has carried out researches both in the field of biology and physics?
    + Jagdish Chandra Bose
1052. Nobel Alfred Bernhard after whom Nobel prizes are given was
    + Both engineer and chemist
1053. India's first remote sensing satellite (IRS 1A) was launched from
    + Baikonour
1054. Theorphrastus is called the father of
    + Botany
1055. The resources which can be used continuously, year-after-year are called
    + Renewable
1056. Radioactive element which has been found to have large reserves in India is
    + Thorium
1057. Non-conventional source of energy best suited for India is
    + Solar energy
1058. Refrigerators keep food unspoiled because
    + At its low temperature, bacteria and moulds are inactive
1059. Water boils at a lower temperature on the hills because

- There is a decrease in air pressure on the hills
1060. Who gave the first evidence of the Big-Bang Theory?
    - Edwin Hubble
1061. The deepest location on the earth's surface on record is about 11.034 km beneath the sea level. It is located in –
    - Marina Trench, west Pacific Ocean
1062. The scientific name of Indian Tiger is
    - Panthera tigris
1063. German Silver is an alloy of
    - Copper, Zinc and Nickel
1064. Which anti-inflammatory drug used extensively to treat cattle in India is found to be the main culprit of vanishing population of Indian vultures –
    - Diclofenac
1065. What is a malicious technique of tricking web users into revealing confidential information called
    - clickjacking?
1066. Which of the following indigenously built payload carried on board with Chandrayaan-I helped NASA's Moon Mineralogy Mapper (M3) to detect water on moon surface is
    - Hyper Spectral Imager (HySI)
1067. What is the source of electric energy in an artificial satellite?
    - Solar cells
1068. The branch of physiology and medicine concerned with heart is known as
    - Cardiology
1069. When a green leaf is seen in red light, its colour will be
    - Black
1070. Study of life in outer space is
    - Exobiology
1071. Instrument used in aircraft to measure altitude.
    - Altimeter
1072. Instrument to measure intensity of sound.

- Audiometer
1073. Instrument used for measuring quantities of heat.
   - Calorimeter
1074. Instrument used for recording sound under water
   - Hydrophone
1075. Instrument used to record graphically various physiological movements i.e., blood pressure, heart beating, study of lungs etc. in living beings.
   - Kymograph
1076. Instrument by which the distance covered by wheeled vehicles is recorded.
   - Odometer
1077. Instrument for recording electrical or mechanical vibrations.
   - Oscillograph
1078. Apparatus used to compare the illuminating power of two sources of light.
   - Photometer
1079. Instrument for recording high temperatures from a great distance.
   - Pyrometer
1080. Instrument for measuring the emission of radiant energy
   - Radiometer
1081. Apparatus for recording of rainfall at a particular piece.
   - Rain gauge
1082. Instrument to measure refractive indices.
   - Refractometer
1083. Instrument for determining the amount of sugar in a solution.
   - Saccharimeter
1084. An optical instrument used for finding out the altitude of celestial bodies and their angular distances.
   - Sextant
1085. Instrument used for measuring arterial blood pressure.
   - Sphygmomanometer

1086. Instrument for measuring curvature of surfaces.
　　　✦ Spherometer
1087. Instrument for determining speeds of aeroplanes and motor boats.
　　　✦ Tachometer
1088. Instrument used to regulate the temperature to a particular degree.
　　　✦ Thermostat
1089. An electrical apparatus used to convert high voltage to low and vice versa
　　　✦ Transformer
1090. India's first satellite Aryabhatt was launched from which country?
　　　✦ Soviet Union
1091. Electrical heating devices are usually made of
　　　✦ Nichrome
1092. Diseases cause by mercury waste is known as
　　　✦ Minamata
1093. At what decibels does sound become noise pollution?
　　　✦ 80 decibels
1094. Homeopathy was discovered by
　　　✦ Samuel Hahnemann
1095. The first successfully cloned animal is
　　　✦ Sheep
1096. Iodine number is an indication of
　　　✦ Degree of unsaturation
1097. Malarial fever is caused by
　　　✦ Plasmodium
1098. Which carbohydrate is rich in honey?
　　　✦ Fructose
1099. In which part of the body lipid digestion takes place?
　　　✦ Small intestine
1100. The term 'gene' was coined by
　　　✦ Johannson
1101. What is the chemical name of quick lime?

Science Quiz

+ Calcium oxide
1102. Electronvolt is a unit of
   + Energy
1103. A ligament tissue connects
   + Bone to bone
1104. Amalgam is an alloy of
   + Mercury with other metal
1105. Which body part gets affected in Cirrhosis?
   + Liver
1106. Sodium metal is stored in
   + Kerosene
1107. Bilirubin is a
   + Bile pigments
1108. The energy content of food is generally measured in
   + Calories
1109. Wi Max is related to
   + Communication technology
1110. Which types of waves are used in a night vision equipment?
   + Infra-red waves
1111. Very small time intervals are accurately measured by
   + Atomic clocks
1112. The unit used to measure the supersonic speed is
   + Mach
1113. On the day the sun is farthest to the earth, the earth is said to be in
   + Aphelion
1114. Transformer works on the principle of
   + Mutual Induction
1115. Henry is the standard unit of
   + Electrical inductance
1116. Total number of ear bones are
   + 3
1117. The functional unit in ecology is
   + Ecosystem
1118. The drug used in the prevention and control of malaria is

✦ Chloroquine
1119. The first synthetically prepared organic compound was
✦ Urea
1120. Artificial silk is also known as
✦ Rayon
1121. The science of ageing is called
✦ Gerontology
1122. Silk is a product of
✦ Salivary gland of larva
1123. Wound healing is enhanced by
✦ Vitamin C
1124. Life span of human white blood corpuscles is
✦ Less than 10 days
1125. A person with diabetes mellitus does not secrete enough
✦ Insulin
1126. First hormone produced artificially by culturing bacteria is
✦ Insulin
1127. The big bang theory explained the origin of
✦ Universe
1128. The plants receive nitrogen in the form of
✦ Nitrate
1129. Which part of human body has greatest number of sweat glands?
✦ Palm of the hand
1130. Who had started vaccination?
✦ Edward Jenner
1131. Proteins consist of
✦ Amino acids
1132. Liver, milk, egg yolk, fish liver oil is a source of:
✦ Vitamin D
1133. Heart is made up of
✦ Cardiac muscle
1134. Hargobind Khurana's work relates to
✦ Understanding the genetic code
1135. Which acid is secreted in the stomach?

✦ HCl (Hydro chloric acid)
1136. Which vitamin is needed to prevent xerophthalmia?
✦ Vitamin A

Identify the Scientific Instruments and Equipment:

1137. Graduated cylinder

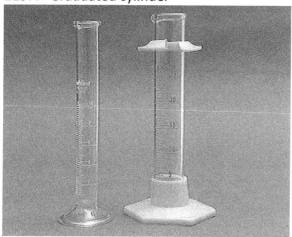

1138. Ring stand

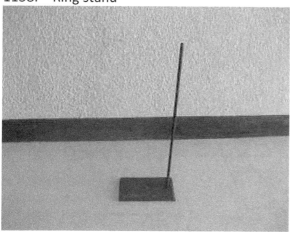

1139. Pipet

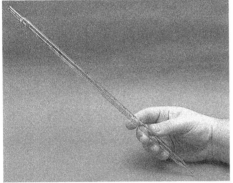

1140. clay triangle

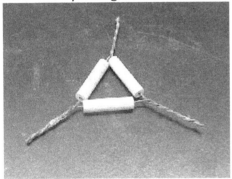

1141. test tube holder

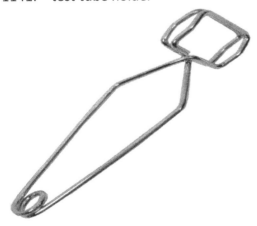

1142. evaporating dish

1143. stir rod

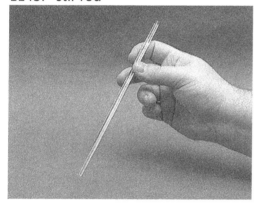

1144. Wire gauze

1145. Utility clamp

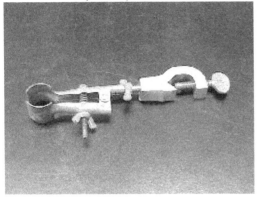

1146. Florence flask

1147. Beaker

1148. wash bottle

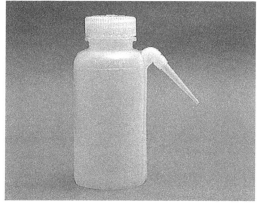

1149. Beaker tongs

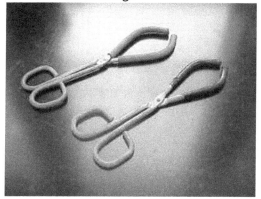

1150. Erlenmeyer flask

1151. Bunsen burner

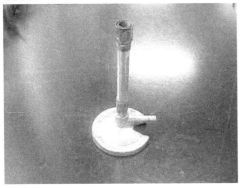

1152. Funnel

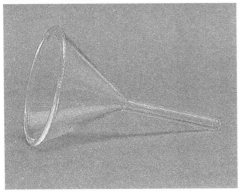

1153. Spatulas

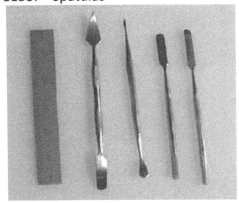

1154. Crucible

1155. Test tube

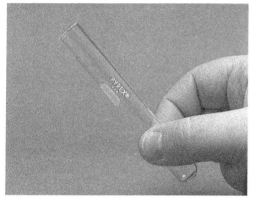

1156. Crucible tongs

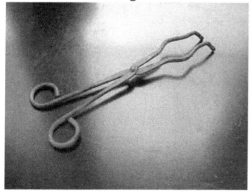

1157. Test tube brushes

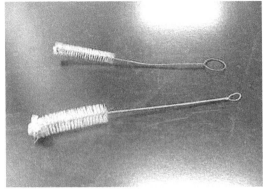

1158. Iron ring

1159. Glass plate

1160. Forceps

1161. Dropper

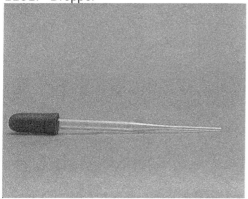

1162. Watch glass

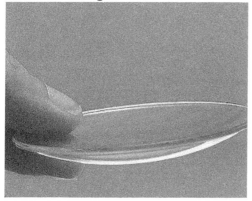

1163.   Test tube rack

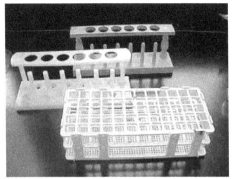

1164.   Mortar and pestle

1165.   Rubber stopper

1166. Filter paper

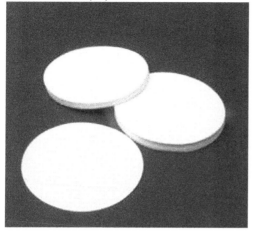

1167. Hot plate / stir plate

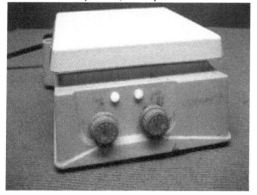

1168. Electronic balance

1169. Thermometer

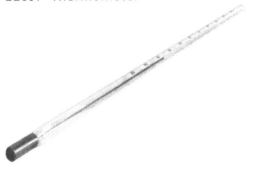

1170. Slides

1171. Microscope

1172. Coverslips

1173. Beam balance

1174. unequal-arm balance

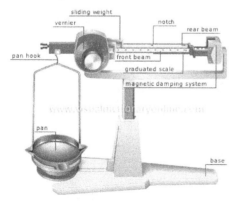

1175. Roberval's balance

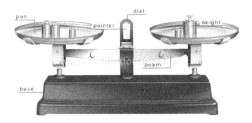

1176. Electronic scale

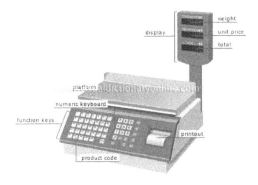

1177. Analytical balance

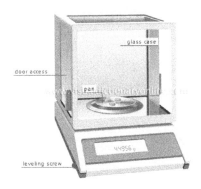

1178. Spring balance

1179. Vernier calliper

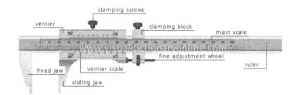

1180. Micrometer calliper

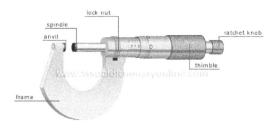

Science Quiz

Famous Scientists:

Science Quiz

Science Quiz

Science Quiz

Science Quiz

CPSIA information can be obtained
at www.ICGtesting.com
Printed in the USA
BVHW041156141122
651890BV00008B/239